Atlas of the Millipedes (Diplopoda) of Britain and Ireland

Dedicated to the memory of

STEVE HOPKIN

who died tragically in a car accident during the production of this work

Atlas
of the
Millipedes (Diplopoda)
of
Britain and Ireland

Paul Lee

British Myriapod and Isopod Study Group

With a contribution from

Paul T. Harding

Photographs by

Steve Hopkin *and* Paul Richards

PENSOFT

Sofia–Moscow

2006

ATLAS OF THE MILLIPEDES (DIPLOPODA) OF BRITAIN AND IRELAND

By Paul Lee

With a contribution from Paul T. Harding

Photographs by Steve Hopkin and Paul Richards

First published 2006

ISBN-10: 954-642-277-0

ISBN-13: 978-954-642-277-4

Pensoft *Series Faunistica* No 59

ISSN 1312-0174

Published by Pensoft on behalf of the Biological Records Centre as part of their series of atlases of the fauna and flora of the UK

http://www.brc.ac.uk

Pensoft Publishers

Geo Milev Str. 13a, Sofia 1111, Bulgaria

pensoft@mbox.infotel.bg

www.pensoft.net

Printed in Bulgaria, September 2006

Contents

Foreword

The publication of the Atlas of the Millipedes of Britain and Ireland less than twenty years after the appearance of the Preliminary Atlas in 1988 has been made possible by the maintenance of a most efficient recording scheme, in combination with the dedication and initiative of the myriapodologists collecting data in the field.

The main participants in this work are mentioned in the chapter in this atlas giving a brief history of millipede recording in Britain and Ireland, written by Paul Harding, himself instrumental in setting up the British Myriapod Survey at the Biological Records Centre at Monks Wood in Cambridgeshire in 1970. I will not detract from his contribution by going into details here. However, it should be said that the recording cards for millipedes, and those for centipedes, were based on those of the original recording system of Harding and Sutton for woodlice. These were intended to collect information on habitat preferences as well as on distribution. From its inception the recording scheme has placed emphasis on collecting important ecological and biogeographical data, an approach which has been much admired far beyond the borders of Great Britain and Ireland. Crucial to this scheme, after the setting up of the Biological Records Centre at Monks Wood, were the initiative of John Heath in organising recording schemes and the foresight and ecological expertise of Paul Harding and Colin Fairhurst when putting in place the Isopod and Myriapod surveys.

At this time Gordon Blower was already a monumental figure. Not only had he assembled all the British and Irish data and published the invaluable Linnean Society millipede volume with keys, but he had encouraged and stimulated a number of us to study and collect information on millipedes. He continued to do this throughout his professional career at the University of Manchester. Moreover, he initiated the careers of a number of excellent research students. Both his scholarship and his meticulous attention to detail were an inspiration to all of us who were working with millipedes in these islands – and beyond. A fact recognised by his election as a member of the Council of the International Centre of Myriapodology. His contribution to our knowledge of the millipede fauna and his catalytic effect with regard to field studies were of immense value. Gordon was an excellent teacher, who published some very significant works on millipedes and had a well-deserved international reputation.

Encouraged by the publication of Gordon's easily workable keys and by the British Myriapod Group Bulletin showing the distribution of millipedes by vice-counties, a number of immensely energetic collectors sprang into prominence. Among these were the successive recording scheme organisers, Colin Fairhurst, Douglas Richardson and Dick Jones, who scarcely left a stone unturned in the regions in which they lived. Large though it is, the Watsonian county of Yorkshire was entirely covered by Douglas Richardson and his colleagues. The present organiser, Paul Lee, a Yorkshireman himself, who worked with Douglas on the myriapods of the county for seven years and now lives in Suffolk, succeeded Dick Jones in 1998 and continued with the same vigour. He is to be complimented on bringing this new atlas to fruition.

Another avid collector was Adrian Rundle who was in first place on the list of contributors for the number of records in the Preliminary Atlas in 1988. There have been so very many contributors to the present volume that it is not appropriate to list them all here. However, apart from the people already mentioned, Steve Gregory, who tops the present list with a staggering total of over 5500 records, and Tony Barber, who has been collecting for almost forty years and has accumulated about 3000, fully deserve a mention. Also, Roy Anderson, Martin Cawley and Declan Doogue have made a huge contribution in recording the Irish fauna.

This new volume will be of great value in conservation work, allowing research workers and planners to see in which types of habitat and microhabitat the different species may be found and which of these require to be protected in order to maintain viable populations of the species in which they are interested. An obvious example would be the value of dead wood to many members of the soil fauna. It obviously also permits an assessment of the biodiversity of different geographical areas and biotopes, and indicates to field workers where to look for new records of given species.

A further feature and major strength of this volume is the placing of the British and Irish fauna in a European context, evidently permitting a better understanding of the species' distributions and the relative importance of the British and Irish populations globally. This is a relatively new departure. Hitherto, although there have been notable exceptions, the vast majority of atlases have been either regional or national, paying scant attention to global range. Fortunately the millipede faunas of neighbouring and central European countries have been reasonably, or in some cases very, well studied and so it is possible to make the connection with a high degree of certainty. Most of the British and Irish species are endemic in the Atlantic zone or are found principally in central Europe from where they extend into Scandinavia as well as here. With regard to the Atlantic zone the only areas which are not well documented are northern Portugal, north-western Spain and parts of the west of France, areas to which the attention of the British Myriapod and Isopod group has recently been drawn. This southern part of the Atlantic zone has now been targeted and is being investigated. With the help of the International Centre of Myriapodology at the Natural History Museum in Paris there has been an excellent collaboration with our neighbours. Indeed there has been a great deal of international collaboration in this field for nearly forty years and a European recording scheme in place since 1978.

Because of the attention paid to sound ecological principles and its biogeographical attributes this atlas should appeal to a broader field of readers than the title might indicate. It reveals distribution patterns that are mirrored in other groups of organisms and is a model of its kind.

Desmond Kime

February 2006

Acknowledgements

There are so many people that have played a part in millipede recording in Britain and Ireland over the years and who have contributed to the preparation of this atlas that it is difficult to single out individuals. However, from my own point of view, it is true to say that I would never have been compiling this atlas were it not for the help and encouragement that Doug Richardson freely gave to a young, novice myriapodologist many years ago. I hope that on reading this work he feels it was worth the time and effort he put into stimulating my initial interest.

In the course of preparing this atlas I have received support and encouragement from members of staff at BRC and from many members of the British Myriapod and Isopod Group. Paul Harding must be singled out for thanks for his assistance under both headings. His intervention has been invaluable in requesting assistance from the relevant BRC staff and in the general guidance he provided as the project progressed. I am also grateful for the chapter he has contributed outlining the history of millipede recording in Britain and Ireland. Many thanks are also due to Henry Arnold, Gavin Broad, Val Burton and David Roy at BRC. Henry and Gavin have been very patient and responded quickly to my many requests to query or amend the dataset over the last few years. Val Burton has had the unenviable task of inputting the data from the record cards, something she achieved with an amazing degree of accuracy. David Roy offered essential advice on the analysis of the ecological data. It was he who came up with the idea of a measure of association and set up the worksheets to calculate it although any mistakes in the interpretation are my own. I also thank him for proof reading the data analysis chapter. Of the BMIG members, Dick Jones, Doug Richardson and the late Colin Fairhurst must be thanked for their contributions as recording scheme organisers without which there would have been no dataset to map or analyse; Steve Hopkin and Paul Richards graciously provided some of their excellent photographs free of charge; Gordon Corbet, John Harper and Paul Richards have given helpful feedback on draft species maps and species accounts; Roy Anderson who, at very short notice, supplied his entire set of over 1300 records from Ireland; Helen Read, Tony Barber, Darren Mann, Bjarne Meidell and Henrik Enghoff have responded promptly to requests for assistance with obtaining literature; and Henrik and Des Kime have patiently responded to numerous questions regarding taxonomy and the ecology and distribution of our millipede species in the rest of Europe. My thanks go to all of these and to all other BMIG members who have helped in any way with the production of this volume.

No national distribution atlas can be produced without the help of a large number of recorders willing to commit their free time, firstly to finding the species in the field and secondly to the administrative task of completing and submitting record cards. When the record cards require the amount of detail found on the RA59 card it is all the more credit to those dedicated workers who have supplied hundreds, and in some cases thousands, of records. The list below does not attempt to differentiate between recorders who have submitted one or one thousand records. It is simply a recognition of all those individuals, some long dead, and organisations who in some way have contributed to the dataset that made this volume possible.

Addey, J.E.; Alcock, I.; Alexander, K.; Anderson, R.; Andrew, R.; Armitage, J.S.; Arnold, G.A.; Arnold, V.; Arthur, W.; Askew, M.E.; Bagnall, R.S.; Bailey, J.; Baker, A.N.; Baker, R.; Baldwin, I.; Bamfield, D.; Banwell, P.A.; Barber, A.D.; Barnes, R.M.; Barnes, S.; Barnish, D.; Barrett, B.; Bass, J.; Bates, J.; Beaumont, D.; Beccaloni, G.; Benfield, M.; Bilton, D.T.; Bioscan; Bishop, M.J.; Bland, K.; Bloor, K.P.; Blower, J.G.; Bloxham, M.G.; Boag, B.; Bolton, D.E.; Boyce, D.; Boyd, J.; Brackenbury, A.; Brade-Birks, H.K.; Brade-Birks, S.G.; Branson, R.D.; Bratton, J.H.; Bray, R.; Brinklow, R.; Britt, D.P.; Britten, H.; Brown, J.D.H.; Bull, A.; Burton, P.J.; Butler, D.M.; Butler, P.M.; Butterfield, J.; Buxton, N.; Calvert, A.; Cambridge Museum; Campbell, B.A.; Campbell, J.M.; Campbell, L.; Carlyle, J.; Carr, J.W.; Carter, H.; Cartwright, G.; Cawkwell, R.; Cawley, M.; Challand, B.; Chater, A.O.; Chatfield, J.; Clark, A.; Clarkson, K.; Clements, D.K.; Clinging, R.;

Cloudsley-Thompson, J.L.; Cocks, W.P.; Coleman, R.; Collins, D.; Collis, G.; Comont, J.; Connell, B.M.; Copson, P.J.; Corbet, G.B.; Cornish, C.; Cotton, D.C.F.; Craig, S.; Crawford, A.; Crawshaw, D.L.; Crew, M.; Crittenden, M.; Cross, C.; Cross, S.A.; Crowson, R.A.; D.A.E.S.; Daniel, R.J.; Darlington, J.P.E.C.; Davey, S.R.; Davis, B.N.K.; Davis, J.E.; Dawe, E.; Daws, J.; Dawson, J.E.; Dawson, N.; Dawson, P.; Dawson, R.W.B.; Delaney, M.J.; Dempster, J.P.; Dennis, P.; Denton, M.L.; Dickinson, J.A.; Dickman, C.; Dixon, J.; Dohle, W.; Dolling, M.H. Mrs.; Donwell, D.; Dony, C.M. Mrs.; Doogue, D.; Douglas, W.; Drake, C.M.; Duffey, E.; Dundee Museum; Durant, G.; Eason, E.H.; Edwards, C.A.; Edwards, E.E.; Edwards, J.J.; Edwards, P.St.J.; Elliott, M.; Ellis, E.A.; Ely, W.A.; Emberson, R.M.; English Nature; Evans, I.; Evans, M.; Evans, R.; Evans, W.; Eversham, B.; Fairhurst, C.P.; Felton, C.; Fenton, G.A.; Ferguson, A.S.; Ferguson, D.; Fitton, M.G.; Flanagan, J.; Fogan, M. Mrs.; Fowles, A.P.; Fredrickson, D.; Friends of Museum Expedition; Frost, C.; Furnival, A.D.; Fussey, G.D.; Gabbutt, P.D.; Gallagher, D.; Garland, S.P.; Garner, D.; Garner, R.; Garrad, L.S.; Garside, A.; George, R.S.; Gibson, L.; Gilbert, O.; Goddard, D.G.; Godfrey, A.; Goldsmith, J.; Grainger, E.; Gray, J.R.A.; Green, S.V.; Gregory, S.J.; Grove, L.R.D.; Grove, S.J.; Guntrip, D.W.; Halfpenny, G.; Halfpenny, J.; Halpin, B.; Hambler, C.; Hames, C.A.C.; Hancey, R.; Hancock, G.; Handley, J.; Hansen, D.; Harding, P.T.; Hardman, J.A.; Harper, J.F.; Harris, G.J.; Harris, J.I.; Harris, J.M.; Harvey, P.R.; Hawkins, K.M.; Hazlehurst, J.; Heaver, D.; Henderson, A.; Higgins, D.; Hill, K.; Hills, C.C.; Hitch, C.J.B.; Hodges, M.; Holmes, P.; Holmes, R.A.; Hooper, H.; Hopkin, S.P.; Hopkins, I.J.; Hughes, M.R.; Hunnisett, J.; Huntley, M.P.; Hyman, P.S.; I.T.E.; Ireland, M.;Irving, D.; Irwin, A.G.; Ivimey-Cook, P.; Jackson, A.R.; Jackson, N.C.S.; Jackson, W.T.; Jacob, R.B.; Jameson, H.L.; Janes, L.; Janes, N.; Jardine, N.; Jefferies, D.K.; Jeffries, H.G.; Jenkins, P.; Jones, B.J.H.; Jones, B.M.; Jones, D.; Jones, G.H.; Jones, J.M.M.; Jones, P.E.; Jones, R.E.; Jones-Walters, L.; Judd, S.; Keay, A.N.; Kendall, P.; Kennard, G.D.; Kennaugh, J.; Key, R.S.; Kime, R.D.; Kirby, P.; Kirkpatrick, D.; Lambert, S.J.; Lambley, P.W.; Langton, P.H.; Lass, R.A.; Lavender, A.; Law, N.; Lazenby, A.S.; Le Quesne, W.; Leach, W.E.; Lee, E.E.; Lee, J.; Lee, P.; Legg, G.; Lewis, J.G.E.; Limbert, M.; Liverpool Museum Survey Team; Lloyd-Evans, L.; Lock, L.; Lott, D.A.;

Loxton, R.G.; Maclean, D.; Maddocks, K.; Mahon, A.; Mander, P.B.; Manning, S.; Martin, Mrs M.E.R.; Masefield, J.R.B.; Mathers, S.; Matthews, M.; Maude, D.; Maxwell, S.; Maxwell, W.G.R.; McAllister, R.I.; McCann, A.G.; McDermatt; McEvoy, D.; McLean, A.; McMillan, N.F.; McNaughton; McQuirk; McShane, A.; Mearns, C.; Meharg, M.; Meloy, B.J.; Mesibov, R.; Meyer, E.; Mileto, R.; Millatt, W.E.; Miller, K.; Miller, P.F.; Milner, J.E.D.; Milne-Redhead, E.; Mitchell, D.; Moller, G.J.; Moran, S.; Morgan, G.L.; Morgan, I.K.; Morgan, M.J. Mrs.; Morison, C.D.; Morley, C.; Moseley, M.; Mosley, O.; Mothersill, C.; Motley, G.; Mounsey, J.D.; Mudge; Murphy, M.D.; Murphy, R.P.; Musker, G.; Musson, D.; N.C.C. (various surveys); N.E. Fife District Council, Environmental Health Dept.; National Museum of Ireland; National Museum of Wales; Natural History Museum; Nau, B.S.; Nelson, J.M.; Nicoll, M.; Norris, A.; Norwich Castle Museum; O'Connor, J.P.; Oliver, P.G.; Owen, D.; Oxford BRC; Parker, A.; Parr, M.J.; Parsons, A.; Parsons, E.; Paul, C.R.C.; Payne, J.; Pearce, E.J.; Pearson and White; Philp, B.; Philp, E.G.; Pirie, M.; Pitt, R.; Pittkin, D.M. Mrs.; Plant, C.W.; Plant, R.A.; Pocock, R.I.; Ponpard, E.; Poole, E.T.; Porter, C.; Porter, D.; Pratley, P.; Pressland, K.O.; Pulman, A.G.; Purshouse, A.; Purshouse, E.; Radley, G.; Rands, D.G.; Rands, E.B.; Rawcliffe, C.P.; Read, H.J.; Read, M.; Read, R.W.J.; Rear, D.; Reardon, N.; Reavell, P.E.; Redgate, N.; Redshaw, E.J.; Reimer, A.; Rhodes, J.D.; Richards, J.P.; Richards, J.P. Jnr.; Richards, O.W.; Richardson, D.T.; Riley, T.H.; Rixom, W.G.; Roberts, C.J.; Robertson, I.; Robertson, J.; Rogers, S.; Rolfe, S.W.; Rowe, G.; RSPB; Rundle, A.J.; Rundle, K.M. Mrs.; Russell, R.; Ruth, E.; Sakwa, N.; Samouelle; Sander, J.J.; Sankey, J.H.P.; Saville, B.; Sayer, K.; Schatz, H.; Scott-Langley, D.; Scowen, A.P.; Selbie, C.M.; Selden, P.; Sewell; Shardlow, M.; Shaw, M.; Shaw, R.; Side, K.C.; Silliprandi, K.; Simmons, M.; Skidmore, P.; Slack, L.; Slawson, G.C.; Smith, C.J.; Smith, D.H.; Smith, E.H.; Smith, E.T.; Smith, R.W.; Sorby Natural History Society; Southward, E.C.; Spalding, D.; Speck, K.; Speight, M.C.D.; Spencer, G.G.; Spirit, M.G.; Spriggs, A.I.; Staffs Nat. Cons. Trust; Stebbings, R.E.; Stelfox, A.W.; Stenton, P.G.; Stevenson, B.; Stevenson, R.; Steward, D.I.; Stewart, J.; Stokes, F.G.; Stones, A.; Stubbs, A.E.; Suffolk Wildlife Trust; Sullivan, C.; Summer, A.; Sutton, C.A.; Sutton, D.; Sutton, S.L.; Taplin, J.; Taylor, A.W.; Taylor, R.T.; Teagle, W.G.; Terrell-Nield, C.E.;

Terry, J.; Tew, G.S.; Thomas, R.; Thomas, T.J.; Thomas, W.B.; Thompson, H.W.; Tilling, S.; Tinning, P.C.; Toher, S.J.; Topley, P.; Tosh, G.; Townsend, M.; Trett, J.; Trett, R.; Trew, A.; Turk, F.A.; Turner, A.; Turner, P.F.; Twigg, D.W.; Twissell, C.; Varndell, I.M.; Varty-Smith, J.C.; Vine, A.E.; Vizzard, D.; Walden, H.W.; Walker, J.; Walker, P.; Wallace, B.; Wallace, I.D.; Walton, J.; Wardaugh, M.; Warmingham, S.; Watson, S.; Watt, K.; Welch, R.C.; Wells, R.; Wells, T.C.E.; Weston, M.R.; Whiteley, D.; Whitmore, M.; Whitton, P.W.; Wilde, I.; Wildlife Trusts; Wilmore, A.; Wilson, F.M.; Wilson, G.; Wilson, J.; Wilson, R.; Wolfe, M.; Woolmer, F.; Wormell, P.; Worthington, E.B.; Wright, I.R.; Wylie, A.; Wytham Survey; Yalden, D.W.; Yeates, C.S.V.; Zordo, I. de.

Photographic credits

Steve Hopkin: front cover (*Blaniulus guttulatus*), pp. 44 (*G. marginata* rolled), 48, 50, 52, 56, 60, 78, 82, 102, 134, 138 (*C. punctatus* walking), 150, 152, 156 (*Paraspirobolus lucifugus*).

Paul Richards: pp. 42, 44 (*G. marginata* unrolled), 74, 86, 88, 94, 96, 104, 108, 116, 130, 138 (*C. punctatus* spiralled), 142, 156 (*Poratia digitata*).

Life histories

The life histories of many of the British and Irish millipedes have been examined in detail but there remain some species whose life histories are totally unknown. From those species which have been studied it is clear that our millipedes exhibit a very varied range of life histories. The majority lay their eggs in the spring or early summer although egg laying in the winter-active Chordeumatidae occurs earlier in the year. Once the eggs have hatched the juvenile millipedes undergo a series of moults at which a variable number of extra segments are added to the body. The number of moults and the time taken to reach maturity varies between taxa but can also differ for individuals of the same species. It may take less than a year for *Brachydesmus superus* to mature (Stephenson, 1961) but up to four years for *Allajulus nitidus* (Blower and Miller, 1977).

Once mature both sexes may either breed just once and then die or they may survive to breed in several seasons; females of *Glomeris marginata* may live for ten years or more. The first strategy is seen in annual species such as *Brachydesmus superus* and in species which take longer to mature such as *Ophyiulus pilosus*. Where breeding occurs over several seasons the males may die after their first pairing so only the females will breed more than once. This strategy is seen in a number of our millipedes including *Cylindroiulus punctatus* and *Polydesmus angustus.* Where males survive for several seasons the adult males of some species are able to moult into sexually non-functional animals. A further moult is then required before the millipede becomes a functional male again. This type of life history is demonstrated by *Tachypodoiulus niger*.

A few of our species are parthenogenetic, the females are able to produce further female offspring from unfertilised eggs. Specimens of *Proteroiulus fuscus, Nemasoma varicorne* and the bristly millipede *Polyxenus lagurus* are usually female when they are collected in Britain or Ireland. Males are sometimes found but only very rarely and they are probably non-functional (Enghoff, 1978).

Collecting and recording millipedes

Although millipedes can be taken in pitfall traps and using other passive collecting techniques, the most commonly used methods for searching in a habitat are turning over stones, logs and other debris and grubbing in leaf litter. Suction sampling can be an effective collection technique on short, dry grassland or heath. Beating of trees regularly produces specimens of *Tachypodoiulus niger*. Soil and litter species may be collected by sieving and sorting of substrate samples or the use of extraction funnels. Larger, more active species can be picked up carefully with fingers or fine forceps. Smaller species are best collected with an entomological pooter or the wetted tip of a fine paintbrush or finger. Specimens are normally transferred directly to tubes of 70% IMS or isopropyl alcohol for transportation back to the laboratory. Specimens should be kept totally immersed in alcohol when being examined under a stereo microscope. Once the species has been determined a specimen can be stored indefinitely in 70% alcohol and should be fully labelled with details of the locality and habitat from which the specimen was taken. Labels should be written on good quality paper using pencil or a waterproof pigment ink. Information on the label should include species name, date of collection, name of collector, name of determiner, name of location specimen was collected at, OS grid reference of location of capture, type of habitat specimen was taken from, method used to collect specimen and any other habitat details such as geology or soil type.

Identification and keys

Identification of many of the British and Irish millipedes can be undertaken in the field with a good handlens once some experience has been gained. In order to gain that experience and to confirm identification of the more difficult species, specimens will need to be examined under a stereo microscope. Magnifications of 30x will be more than adequate for most purposes but up to 75x magnification can be useful when working with some of the smallest species. The most comprehensive guide to the British and Irish fauna is the Linnean Society Synopsis by Blower (1985). Although now out of print it is still possible to obtain copies through specialist entomological book sellers. Nine species found in Britain and one found in Ireland are not described by Blower (1985). Most of these species have been described in the Bulletin of the British Myriapod Group and its successor, the Bulletin of the British Myriapod and Isopod Group. A revision of the Linnean Society Synopsis is in preparation and will include all of the additional species. All of the British and Irish species are covered in a draft key produced by the recording scheme organiser for the benefit of delegates attending the Field Studies Council course on Centipede, Millipede and Woodlouse identification which he leads at the Flatford Mill Field Study Centre each year. Copies of the key can be obtained from the scheme organiser for the cost of photocopying the document.

Where a recorder has any doubts about an identification a voucher specimen should be submitted to the recording scheme organiser for confirmation. The scheme organiser may also request a voucher where an individual record seems surprising based on the known ecology and distribution of the species involved.

Biological records of millipedes, requests for copies of the draft millipede key and requests for assistance with identifications should be sent to:

Paul Lee
Oakdene, The Heath
Tattingstone
Ipswich
IP9 2LX

e-mail: paullee@arachne.fsnet.co.uk

Millipede recording in Britain and Ireland

Paul T. Harding

Growth of the species list (1814 to 2005)

The number of species of millipede known from Britain and Ireland has increased progressively since the first list was published by Leach in 1814 (Figure 1). Growth of a combined list for Britain and Ireland was slow throughout the 19th Century, reaching only 16 species by 1893 (Pocock, 1893). By 1907, Pocock and Evans had raised the total to 23 and by 1922 Bagnall and Brade-Birks had added a further 16 valid species. However, only one further species had been added by 1939 (Brade-Birks, 1939), making a total of 40. The first Linnean Society *Synopsis* (Blower, 1958) included 43 valid species, and by 1969 the total had reached 47 (Blower, 1972b). The revised Linnean Society *Synopsis* (Blower, 1985) included 52 species and two species were added in the *Preliminary Atlas* (British Myriapod Group, 1988). By 2002, the list had grown to 60 species and a further species was collected from hothouses at Kew Gardens in 2004. Another exotic species was identified from the hot-houses of the Eden Project in 2005 bringing the total to 62 species.

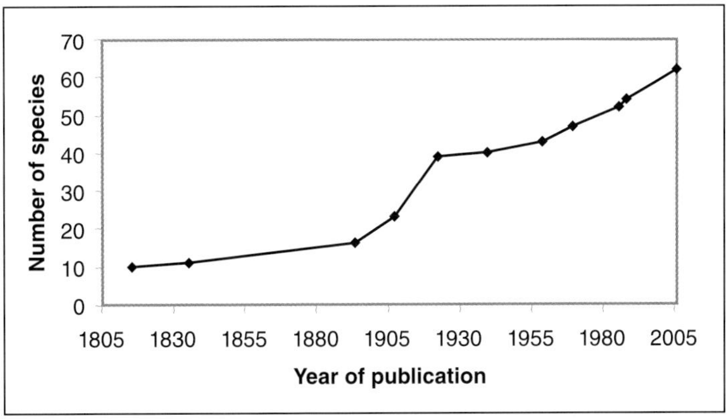

Figure 1: Growth in the number of valid millipede species recorded in Britain and Ireland (1814 to 2005).

Recording millipedes – 1814 to 1969

1814 to 1893

Leach (1814; 1815) provided a baseline, recording ten valid species from Britain. Subsequent lists for Scotland (Johnston, 1835) and Ireland (Templeton, 1836) added only one new species, but both authors identified localities for species. Growth of a combined list for Britain and Ireland was slow throughout the 19th century, but Pocock (1893) suggested that there must be some 50 species of myriapod known to occur in England, "but not one half of these are in print", implying that although species were known to occur, that knowledge was unpublished.

1894 to 1939

Pocock and William Evans published a few papers between 1894 and 1907, bringing forward a further seven species to the list. Then, two experienced naturalists from the north of England began to take an interest in millipedes. Richard Bagnall added ten species between 1912 and 1922; a bibliography of his papers on myriapods was compiled by Barber and Mann (2004). Although A. Randall Jackson may not have added species to the list, he published many distribution records. Perhaps more importantly, he fostered an interest in millipedes with Graham Birks and Hilda Brade, who later married with the surname Brade-Birks. Between 1916 and 1939, they published 36 separate *Notes on Myriapoda* in several journals (Barber & Blower, 1992), including many distribution records, eight species new to the list, three new to science and, finally, a complete checklist (Brade-Birks, 1939).

1940 to 1958

Barber (2003) noted that "*1950 saw the beginning of a new era in myriapod studies in Britain.*" For millipedes, this centred mainly on J. Gordon Blower. He published the first key to British millipedes (Blower, 1952), which was later expanded into a full Linnean Society *Synopsis* (Blower, 1958). These publications made it possible, for the first time, to identify British and Irish millipedes with a reliable, English language key. Seven species had been added to the list prior to the publication of the *Synopsis*, and Blower took the opportunity of its publication to remove six invalid species from the list by Brade-Birks (1939). The *Synopsis* included a summary of the known distribution of each species.

J. Gordon Blower (1923–2001)

Blower's essentially important contribution to the study of millipedes in Britain and Ireland has been described by Barber (2003) and Richards (2003), and a bibliography was included with his obituary (Read, 2002). In 1952, Blower began to publish papers dealing with the identification and occurrence of millipedes and these continued to 1989. His two Linnean Society *Synopses* to the millipedes (Blower, 1958, 1985) have been central to the accurate identification of millipedes for nearly 50 years, backed-up by his personal help to many recorders with checking or identifying difficult specimens. From his base at the University of Manchester, Blower fostered a life-long interest in myriapods among several of his students, in particular, Colin Fairhurst and Helen Read.

1959 to 1969

The impact of these keys (Blower, 1952, 1958) on millipede recording was gradual, but by the late 1960s several authors had begun to publish new records. A small group of active British myriapodologists attended the First International Congress of Myriapodology, held in France in 1968. Blower continued to collate information on the vice-county distribution of species in Britain and Ireland, but this was not published until 1972 (Blower, 1972b). This publication is a remarkable summary and analysis of the available information up to 1969 and includes a comprehensive bibliography of the sources used. An academic summary of this information, updated to 1970, was published by Blower (1974).

Beginning to organise millipede recording – 1970 to 1984

In April 1970, twelve myriapodologists met in North Devon "*with the express purpose of looking for myriapods and talking about them*"

(Blower, 1972a) and the British Myriapod Group (BMG) was formed. In his report on the meeting, Blower noted the following: "*In particular we discussed distribution. Most existing British [and Irish] records are for counties or vice-counties, but Desmond Kime and Tony Barber had already made most of their recording in grid squares. Colin Fairhurst reported on the latest moves to establish a British Myriapod Survey on these lines (Fairhurst & Barber, 1972).*" Volume 1 of the new journal, the *Bulletin of the British Myriapod Group*, was published in April 1975.

British Myriapod Survey

The establishment of the British Myriapod Survey had been under discussion since 1969 and was developed jointly with the Isopod Survey Scheme of the British Isopoda Study Group (BISG). The separately organised surveys of millipedes, centipedes and woodlice would use a single hierarchical habitat recording system on species-list record cards printed for BMG and BISG by the Biological Records Centre (BRC). BMG and BISG had conducted lengthy negotiations with the BRC concerning this habitat recording system, which was the first to be used by any national species recording scheme. Full accounts of the development of the habitat recording system and the development of the respective recording schemes have been given by Fairhurst and Barber (1972), Barber and Fairhurst (1974) and Harding and Sutton (1985). The original millipede record card (RA13) (Figure 2) was produced by the BRC for BMG in June 1970 and was in general use until the end of 1984 (see *Record cards and habitat recording systems* below).

British Myriapod Group

Following a second field meeting, at Kington, Herefordshire in 1971, BMG activities focused on the individual recording schemes and no further meetings were held until April 1982. Following this meeting, Douglas Richardson produced the first BMG Newsletter. A small meeting hosted by the BRC in early April 1983 helped to plan a way forward with both myriapod schemes. It was also suggested that the *Bulletin of the British Myriapod Group* should be revived. Later in April 1983, a joint field meeting with BISG brought together the two groups and the three recording schemes, for the first time.

Establishing the millipede scheme

Colin Fairhurst continued as the national organiser for the millipede scheme, collating records and identifying specimens until about 1984, overlapping with his successor, Douglas Richardson (see below). Fairhurst's involvement with the scheme had also involved several phases of analysis of the data (Fairhurst, Harding & Sutton, 1975; Fairhurst, Barber & Armitage, 1978; Fairhurst & Armitage, 1979; Fairhurst, 1984, 1985). For the joint BMG/BISG field meeting at Lancaster in April 1983, Fairhurst prepared an analysis of habitat data and 50km square distribution maps based on 10,000 records (Fairhurst, 1984). Despite its title, Irish records were included in this report. Only a few copies of this report were circulated. The report disappointed some recorders because of the spatial scale of the distribution maps (50km squares) and because some important recent records had been omitted. Nevertheless, it has been the most comprehensive analysis of habitat data from the millipede scheme until this *Atlas*. In 1975, Fairhurst updated the vice-county distribution tables from Blower (1972b) and further updates were included in the BMG Newsletter in 1982 and 1983, in the *Preliminary Atlas* (British Myriapod Group, 1988) and in the *Bulletin* in 1993 (Jones, 1993).

Organised millipede recording –1985 to 2005

1985 marked a turning point for the millipede scheme from several perspectives, in particular, a new scheme organiser, a new record card, a new identification key, a regular BMG Newsletter and the revival of the BMG *Bulletin*.

Scheme organisers

Douglas Richardson, an enthusiastic field recorder actively involved with the Yorkshire Naturalists Union (YNU), gradually took over as scheme organiser during 1984, a role that he occupied until April 1988. He was succeeded by Dick Jones who, despite ill health, continued as scheme organiser for 10 years. Paul Lee succeeded Dick Jones as scheme organiser in May 1998. Paul Lee had been introduced to millipedes by Douglas Richardson at a YNU residential weekend meeting in

1984 and he is still the YNU's recorder for Other Arthropods, despite living in Suffolk since 1990.

Newsletter

Douglas Richardson initiated a Newsletter for the British Myriapod Group in 1983. By 1985, it had become an established mechanism for communicating news and information among the Group. The BRC produces and mails-out copies to all members of the Group. The role of BMG Newsletter editor seems to follow that of organiser of the Millipede Scheme, passing to Dick Jones in 1988 and to Paul Lee in 1998. The last BMG Newsletter was No 32 (Spring 2000). Its successor, British Myriapod and Isopod Group Newsletter, also edited by Paul Lee, is produced twice a year.

New *Synopsis*

In 1985, the long-awaited Linnean Society *Synopsis* (Blower, 1985) was published. Covering 53 species and with excellent figures, it provided a much needed impetus to millipede recording and has remained the standard work for the last 20 years.

New millipede record card

1985 also saw the introduction of a new millipede record card (RA59) (Figure 3), which had been produced by the BRC for BMG in December 1984. It used the checklist of 53 species, as published in Blower (1985), and introduced a revised habitat recording system (see *Record cards and habitat recording systems* below).

Bulletin of the British Myriapod Group

Plans to revive the *Bulletin of the British Myriapod Group* were prompted by a passing remark at the BRC/BMG meeting in April 1983. A bookseller was reported as offering Volume 1 of the *Bulletin* at a grossly inflated price, as "*the only Volume published*". Gordon Blower rose to this implied challenge and undertook to edit the *Bulletin*; Volume 2 was published in January 1985. Since then, the *Bulletin* has appeared almost annually, usually in time for the annual field meeting. Tony Barber joined Blower as joint editor for Volume 3 onwards, and

Helen Read joined the editorial team from Volume 10 onwards. The Bulletin accepts all types of papers, and some shorter notes, of potential interest to myriapodologists in Britain and Ireland. All papers are refereed. The change to *Bulletin of the British Myriapod and Isopod Group* from Volume 17 onwards has merely extended the taxonomic scope of the journal. The *Bulletin* is self financing and, as BMIG is not a subscription society, is not linked to a membership fee.

Preliminary Atlas of Millipedes

Douglas Richardson and Paul Harding worked with other BMG members and staff at the BRC to publish a *Preliminary Atlas* of millipedes (British Myriapod Group, 1988) including 10km square distribution maps for 47 species and updated vice-county tables. This *Atlas* did not include analyses of the habitat data from the scheme.

Colin Fairhurst (1942–1994)

Colin Fairhurst had played a key role in setting up and running the millipede recording scheme throughout the 1970s and in developing the analysis and interpretation of data for the millipede, centipede and woodlice schemes. In a short paper (Fairhurst, 1985), he summarised his main work with the millipede scheme, but the pressures of his academic career, combined with increasing ill-health, prevented him from taking an active role after 1985.

Ireland

In Ireland, Declan Doogue had been recording millipedes since about 1978 and he collaborated with other BMG members to publish a review of Irish millipedes (Doogue, *et al.* (1993). This review referred briefly to uncertainties about some Irish records, many of which are supported by specimens in the collection at the National Museum of Ireland, Dublin, but Colin Fairhurst's interpretation of his re-examination of this collection was not completed when he died in 1994.

European Myriapod Survey

As early as 1978, the 4th International Congress of Myriapodology adopted a proposal for a Europe-wide scheme to record the distribution of myriapods (Kime, 1985). Fairhurst had described an outline of this scheme in an unpublished paper given at the 4th International Colloquium of the European Invertebrate Survey, at Saarbrücken, Germany, in 1979. One of the main intentions had been to collate records using cards with a unified, Europe-wide, habitat recording system. The RA59 card was the only manifestation of this system for millipedes. This highly ambitious plan soon settled into achievable reality under the leadership of Desmond Kime who has collated records from many sources, including his own extensive field surveys. It has resulted in published maps of the European distribution of selected millipede species (Kime, 1990, 1999, 2001), with many more maps still awaiting publication.

Annual field meetings

The success of the first combined BMG/BISG field meeting in 1983 has led to a meeting being held each year, normally after Easter, somewhere in Britain or Ireland. Larger numbers of recorders are attracted to combined meetings, normally making-up sufficient numbers to book a residential venue such as a field centre or a university or college hall of residence. Field meetings serve many purposes, but usually venues have been chosen to enable recording in hitherto under-worked areas. Summaries of millipede records from most field meetings have been published in the BMG Newsletter or the *Bulletin* (Table 1). Two overseas expeditions, to Hungary and north-western Spain, have also been organised.

British Myriapod and Isopod Group

Informal discussions, held over several years, resulted in the British Myriapod Group and the British Isopoda Study Group merging to form the British Myriapod and Isopod Group (BMIG). This merger was agreed in 2000 by those attending the annual field meeting at Saffron Walden. Within BMIG, the recording schemes, BMIG newsletter and the renamed *Bulletin of the British Myriapod and Isopod Group* are organised much as before.

Table 1: Field meetings of the British Myriapod Group (1970–1999) and British Myriapod and Isopod Group (2000–2005).

Year	Venue	Grid ref.	Millipede records summarised
1970	Brendon, Devon	SS74	Bulletin BMG, **1**, 1972
1971	Knighton, Herefordshire	SO27	Bulletin BMG, **1**, 1972
1982	Plymouth, Devon	SX45	BMG Newsletter, 1983
1983	Lancaster	SD46	BMG Newsletter, November 1983
1984	Brancaster, Norfolk	TF74	BMG Newsletter, October 1984
1985	Bangor, North Wales	SH57	BMG Newsletter, July 1985
1986	Manchester Polytechnic	SJ89	BMG Newsletter, **6**, Sept 1986
1987	Langford Court, Somerset	ST46	BMG Newsletter, **7**, Sept 1987
1988	Newbridge-on-Wye, Powys	SO05	BMG Newsletter, **9**, Sept 1988
1989	Hallsannery, Devon	SS21	BMG Newsletter, **11**, Sept 1989
1990	Thornham Magna, Suffolk	TM17	BMG Newsletter, **13**, Sept 1990
1991	Swanage, Dorset	SZ07	BMG Newsletter, **15**, Sept 1991
1992	Littledean, Gloucestershire	SO61	BMG Newsletter, **17**, Sept 1992
1993	Hassocks, Sussex	TQ31	BMG Newsletter, **19**, Sept 1993
1994	Meigle, Perthshire	NO24	BMG Newsletter, **21**, Autumn 1994
1994 (May)	Hungary		Bulletin BMIG, **21**, 2006
1995	Rowrah, Cumbria	NY01	BMG Newsletter, **23**, Autumn 1995
1996	Kingcombe, Dorset	SY59	BMG Newsletter, **25**, Autumn 1996
1997	Dalry, Galloway	NX68	BMG Newsletter, **31**, Autumn 1999
1998	Chyvarloe, Cornwall	SW62	BMG Newsletter, **29**, Autumn 1998 & **30**, Spring 1999
1999	Ford Castle, Northumberland	NT93	Bulletin BMIG, **21**, 2006
2000	Saffron Walden, Essex	TL53	
2001	Dingle, Ireland	V59	Bulletin BMIG, **18**, 2002
2002	Youlgreave, Derbyshire	SK26	Bulletin BMIG, **20**, 2004
2003	Reaseheath, Cheshire	SJ65	Bulletin BMIG, **21**, 2006
2004 (March)	Galicia, Spain		
2004 (April)	Aston Clinton, Buckinghamshire	SP81	Bulletin BMIG, **21**, 2006
2005	Durham	NZ24	Bulletin BMIG, **21**, 2006

2003 Conference

Following the recent deaths of two internationally respected British myriapodologists (Ted Eason (1915–99) and Gordon Blower (1923–2001),

Helen Read organised a one-day international conference, hosted by BMIG and Manchester University, to celebrate their scientific and personal contributions. Papers from the conference were published in *BMIG Bulletin* **19**, 2003.

BMIG website (www.bmig.org.uk)

An early innovation by the new Group was the development of the BMIG website by Craig Slawson. As well as providing basic information about the Group, its activities and its publications (including contents lists for the Bulletin), it includes details of forthcoming meetings and changes to check lists.

Record cards and habitat recording systems

Initiating the recording schemes

The Biological Records Centre (BRC) was set up in 1964 and soon became a focal point for many species recording projects. The scope of the BRC expanded rapidly, especially with the recruitment of a zoologist, John Heath, in 1967 (Harding & Sheail, 1992). John Heath had previously worked on a project with Gordon Blower, so it was natural for him to approach Blower to set-up a recording scheme for millipedes. Although Blower declined this invitation, he passed on the proposal to a former student, Colin Fairhurst. Initially, the BRC had proposed a single scheme covering millipedes, centipedes and woodlice, but a scheme covering woodlice had already started, in 1968, independently of the BRC, using a simple, but entirely original habitat recording system designed to collect information about the habitat preferences of woodlice (Harding & Sutton, 1985). As all three taxonomic groups were almost unknown ecologically, this habitat-based approach to species recording was seen to be appropriate for all three groups. The BRC had originally opposed the inclusion of a habitat recording system on record cards, but the aspiring scheme organisers for the three groups eventually reached agreement with the BRC about the scope of three separate schemes, linked by their use of the same habitat recording system. The main constraints imposed by the BRC on the three schemes were that the habitat recording system should be coded for computerisation using available 'space' in the BRC's standard 80-column punched cards and that the schemes should be responsible for analysing the habitat data. Cards for each of the three schemes were printed by the BRC in the summer of 1970.

RA13 recording card (Figure 2)

The RA13 card for millipedes includes the same hierarchical, habitat recording system as the cards for centipedes (RA14) and woodlice (RA15). Inevitably, the habitat system on the cards was constrained by factors such as existing knowledge of the habitat preferences of individual species in all three groups, the amount of unused 'space' on the BRC's 80-column punched cards and the expected willingness, or otherwise, of recorders to complete complicated cards. Results have been published from analyses of the habitat data collected by the surveys of woodlice (Harding & Sutton, 1985) and centipedes (Barber & Keay, 1988). The only previous analyses of habitat data for millipedes were those produced by Fairhurst (1984) for limited circulation.

RA 59 recording card (Figure 3)

The RA59 card was introduced in 1985 to link to the proposed European survey of myriapods. It includes an improved and slightly simplified hierarchical habitat recording system which is the same as that on the RA58 card for centipedes. Although the habitat system used on the RA59 card has many individual fields that are the same as or analogous to fields on the RA13 card, analysis and interpretation of the data has presented several problems.

Check list

The arrangement of taxa in the following check list is based on the classification of the Diplopoda by Hoffmann (1979). The sequences of the orders and of the families of the order Julida have been adjusted to agree with the results of the cladistic analyses by Enghoff (1984, 1991). Species within a genus are arranged in alphabetical order. The nomenclature generally follows that on the Fauna Europaea website (Enghoff & Kime, 2005) and most of the names are those familiar to British and Irish workers from the check list published by Blower (1985) or from the publication of records of species new to the British fauna in the Bulletin of the British Myriapod Group. However, the list contains some minor changes and, where considered necessary, these are outlined in the footnotes below. Only the more recently used synonyms are given.

Class DIPLOPODA
Subclass PENICILLATA
Order POLYXENIDA
Family Polyxenidae
Polyxenus lagurus (Linnaeus, 1758)
Subclass CHILOGNATHA
Superorder ONISCOMORPHA
Order GLOMERIDA
Family Glomeridae
Glomeris marginata (Villers, 1789)
Geoglomeris subterranea Verhoeff, 1908
Stygioglomeris crinata Brölemann, 1913
Geoglomeris jurassica Verhoeff, 1915
Family Doderiidae
Adenomeris gibbosa Mauriès, 1960
Trachysphaera lobata (Ribaut, 1954)

Superorder COLOBOGNATHA

Order POLYZONIIDA

Family Polyzoniidae

Polyzonium germanicum Brandt, 1837

Polyzonium germanicum Brandt, 1831

Superorder NEMATOPHORA

Order CHORDEUMATIDA

Family Craspedosomatidae

Craspedosoma rawlinsii Leach, 1814

Craspedosoma rawlinsii Leach, 1815

Nanogona polydesmoides (Leach, 1814)

Nanogona polydesmoides (Leach, 1815)

Polymicrodon polydesmoides (Leach, 1815)

Family Anthogonidae

Anthogona britannica Gregory, Jones & Mauriès, 1993

Family Chordeumatidae

Chordeuma proximum Ribaut, 1913

Chordeuma sylvestre C.L.Koch, 1847

Melogona gallica (Latzel, 1884)

Melogona scutellaris (Ribaut, 1913)

Melogona scutellare (Ribaut, 1913)

Melogona voigti (Verhoeff, 1899)

Family Anthroleucosomatidae

Anamastigona pulchella (Silvestri, 1898)

Anamastigona pulchellum (Silvestri, 1894)

Family Brachychaeteumatidae

Brachychaeteuma bagnalli Verhoeff, 1911

Brachychaeteuma bradeae (Brölemann & Brade-Birks, 1917)

Brachychaeteuma melanops Brade-Birks & Brade-Birks, 1918

Brachychaeteuma melanops Brade-Birks, 1918

Superorder MEROCHETA

Order POLYDESMIDA

Family Paradoxosomatidae

Oxidus gracilis (C.L.Koch, 1847)

Stosatea italica (Latzel, 1886)

Family Polydesmidae

Brachydesmus superus Latzel, 1884

Polydesmus angustus Latzel, 1884

Polydesmus barberii Latzel, 1889

Polydesmus coriaceus Porat, 1871

Polydesmus coriaceus Porat, 1870

Polydesmus gallicus Latzel, 1884

Polydesmus denticulatus C.L.Koch, 1847

Polydesmus inconstans Latzel, 1884

Propolydesmus testaceus (C.L.Koch, 1847)

Polydesmus testaceus C.L.Koch, 1847

Family Haplodesmidae

Cylindrodesmus hirsutus Pocock, 1889

Cylindrodesmus laniger Schubart, 1945

Prosopodesmus panporus Blower & Rundle, 1980

Family Pyrgodesmidae

Poratia digitata Porat, 1889

Poratia digitata Schubart, 1934

Family Macrosternodesmidae

Macrosternodesmus palicola Brölemann, 1908

Ophiodesmus albonanus (Latzel, 1895)

Superorder JULIFORMIA

Order SPIROBOLIDA

Family Spirobolellidae

Paraspirobolus lucifugus (Gervais, 1836)

Paraspirobolus dictyonotus (Latzel, 1895)

Order JULIDA

Family Blaniulidae

Choneiulus palmatus (Nìmec, 1895)

Nopoiulus kochii (Gervais, 1847)

Proteroiulus fuscus (Am Stein, 1857)

Blaniulus guttulatus (Fabricius, 1798)

Archiboreoiulus pallidus (Brade-Birks, 1920)

Boreoiulus tenuis (Bigler, 1913)

Family Nemasomatidae

Nemasoma varicorne C.L.Koch, 1847

Thalassisobates littoralis (Silvestri, 1903)

Family Julidae

Julus scandinavius Latzel, 1884

Haplopodoiulus spathifer (Brölemann, 1897)

Ophyiulus pilosus (Newport, 1842)

Leptoiulus belgicus (Latzel, 1844)

Leptoiulus kervillei (Brölemann, 1896)

Metaiulus pratensis Blower & Rolfe, 1956

Allajulus nitidus (Verhoeff, 1891)

Cylindroiulus nitidus (Verhoeff, 1891)

Cylindroiulus britannicus (Verhoeff, 1891)

Cylindroiulus caeruleocinctus (Wood, 1864)

Cylindroiulus latestriatus (Curtis, 1845)

Cylindroiulus londinensis (Leach, 1815)

Cylindroiulus parisiorum (Brölemann & Verhoeff, 1896)

Cylindroiulus punctatus (Leach, 1815)

Cylindroiulus salicivorus Verhoeff, 1908

Cylindroiulus truncorum (Silvestri, 1896)

Cylindroiulus vulnerarius (Berlese, 1888)

Enantiulus armatus (Ribaut, 1909)

Unciger foetidus (C.L.Koch, 1838)

Brachyiulus pusillus (Leach, 1815)

Ommatoiulus sabulosus (Linnaeus, 1758)

Tachypodoiulus niger (Leach, 1814)

Tachypodoiulus niger (Leach, 1815)

Notes

1. The Chordeumatid genera *Melogona* and *Anamastigona* are feminine (Jeekel, 1970) and the specific epithets now reflect this. The British and Irish check list has previously contained a mixture of neuter and feminine specific names.
2. There has been confusion over the date of publication by Leach of the six species first described by him. This has partly arisen as his paper was read to the Royal Society in 1814 but was not published

until 1815. Further confusion was caused by references to Leach's manuscript in volume 7 of Brewster's *Edinburgh Encyclopaedia* published in 1814. These references included brief descriptions of *Craspedosoma rawlinsii*, *Nanogona polydesmoides* and *Tachypodoiulus niger* and this earlier publication date is now reflected in the check list.

3. Following cladistic analysis, Djursvoll *et al.* (2001) elevated *Propolydesmus* from a subgenus within Polydesmidae to full generic status. Enghoff and Golovatch (2003) later redefined the genus resulting in the inclusion of *Propolydesmus testaceus*, the only British and Irish species within the new genus.
4. Bagnall (1922) collected a single damaged female specimen of a polydesmid from Gibside, County Durham. He assigned the specimen to *Eumastigonodesmus boncii* (Brölemann, 1908), a species only known elsewhere from the High Pyrenees (Kime, 2001). There is no specimen to support the Gibside record and, despite several attempts to find the species again at the same location (the latest in 2005), it has not been rediscovered. Based on this evidence, Jones (1993) concluded that the occurrence of *Eumastigonodesmus boncii* in Britain must be considered doubtful at best. *Eumastigonodesmus boncii* has therefore been removed from the check list.
5. *Ommatoiulus rutilans* was recorded from Shetland in 2003 (Lee, 2004) when it was found in a consignment of mushrooms imported from Spain. There is no evidence that the species has ever established a breeding population in Britain or Ireland and therefore it has been omitted from the check list.

Distribution maps and species accounts

The distribution maps in this atlas are based on over 47 000 records submitted to the millipede recording scheme over a period of 33 years. Each species map shows the 10km squares from which the species has been recorded, open dots indicating pre-1980 records and solid dots indicating records from1980 onwards. Each dot on a species map may represent a single occurrence of the species at one site within the 10 km square or the continuous existence of numerous populations.

Map 1 is a coverage map showing each 10km square from which there is at least one record of a millipede. A dot on this map may equally result from an intensive survey of an area or a chance encounter with an animal. Similarly the absence of a dot may be due to the absence of survey efforts or the failure of a survey to find any animals. The information on Map 1 is repeated as grey shading on each of the species maps.

Map 2 gives some idea of the intensity of recording of <u>all</u> species within each 10km square. This allows the identification of areas of greatest recording effort. These areas are often centred on the homes of particularly active recorders. The map is useful when interpreting the distribution of a number of uncommon species which show very patchy but widespread distributions. However, this information has not been provided for the individual species.

Map 3 is a species richness map indicating the number of species recorded from each 10km square. Although it can be used to identify possible areas of high biodiversity, the importance of recorder effort in producing species lists must be considered. Therefore this map should only be interpreted in conjunction with maps 1 and 2.

Each of the following accounts begins with a brief description of the British and Irish distribution of the species concerned. This is followed by a summary of the information gathered by analysis of the habitat data from the Millipede Recording Scheme. Where further information on habitat associations has been published this is discussed in relation to the current analysis. Then an attempt has been made to place the geographic distribution of each of the British and Irish species within the context of their European distributions. Kime (2001) recognised seven broad categories of distribution based on climatic zones. The majority of our species can be described as Atlantic, that is they occur only along the western coastal fringes of Europe, including Britain and Ireland, from north west Spain through western France, Belgium, Holland and Denmark to the southernmost tip of Scandinavia. The distribution of Atlantic species is often further restricted by climate within this zone. A significant minority of our species can be considered central European species. These millipedes just reach our shores at the north west edge of their range. Information on the life history of each species is summarised where it is available. A number of accounts also include some details of collecting techniques of particular value for recording the species.

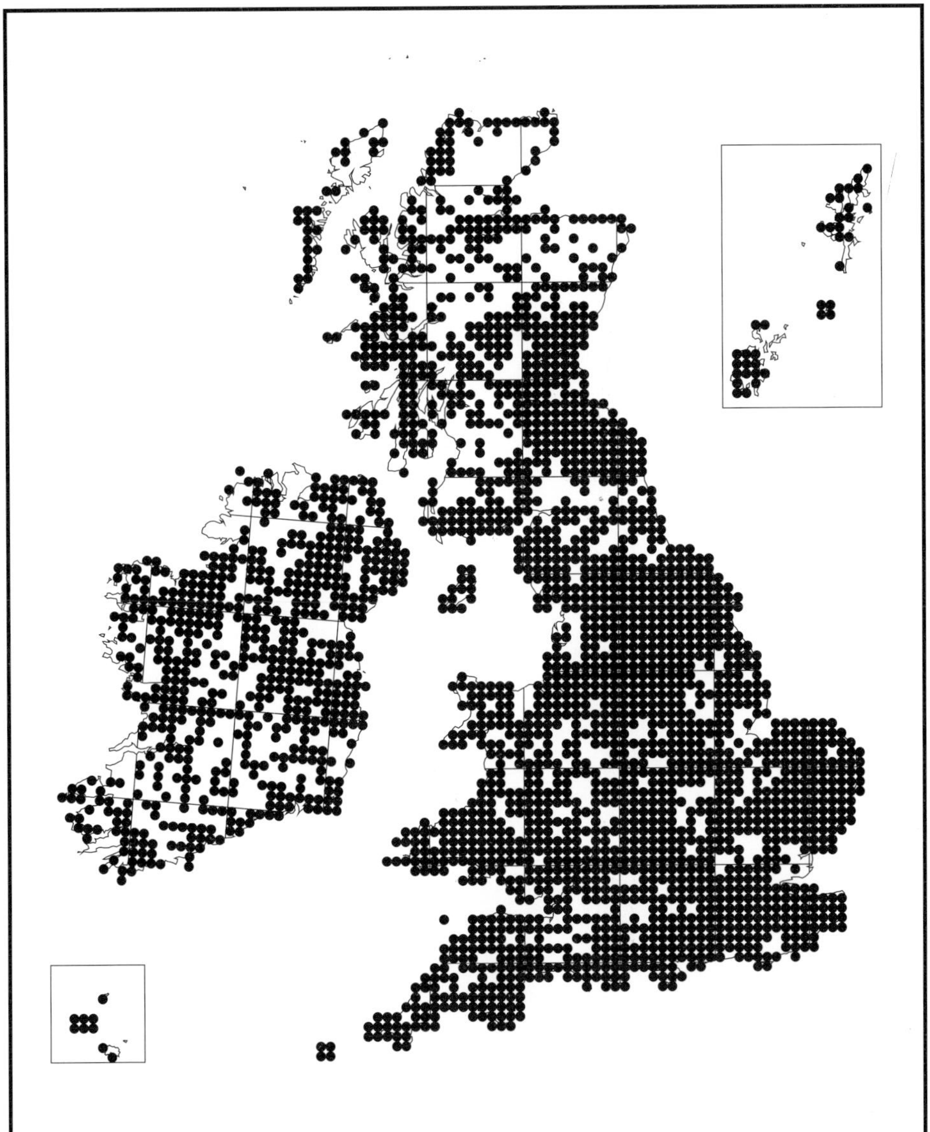

Map 1: Coverage map

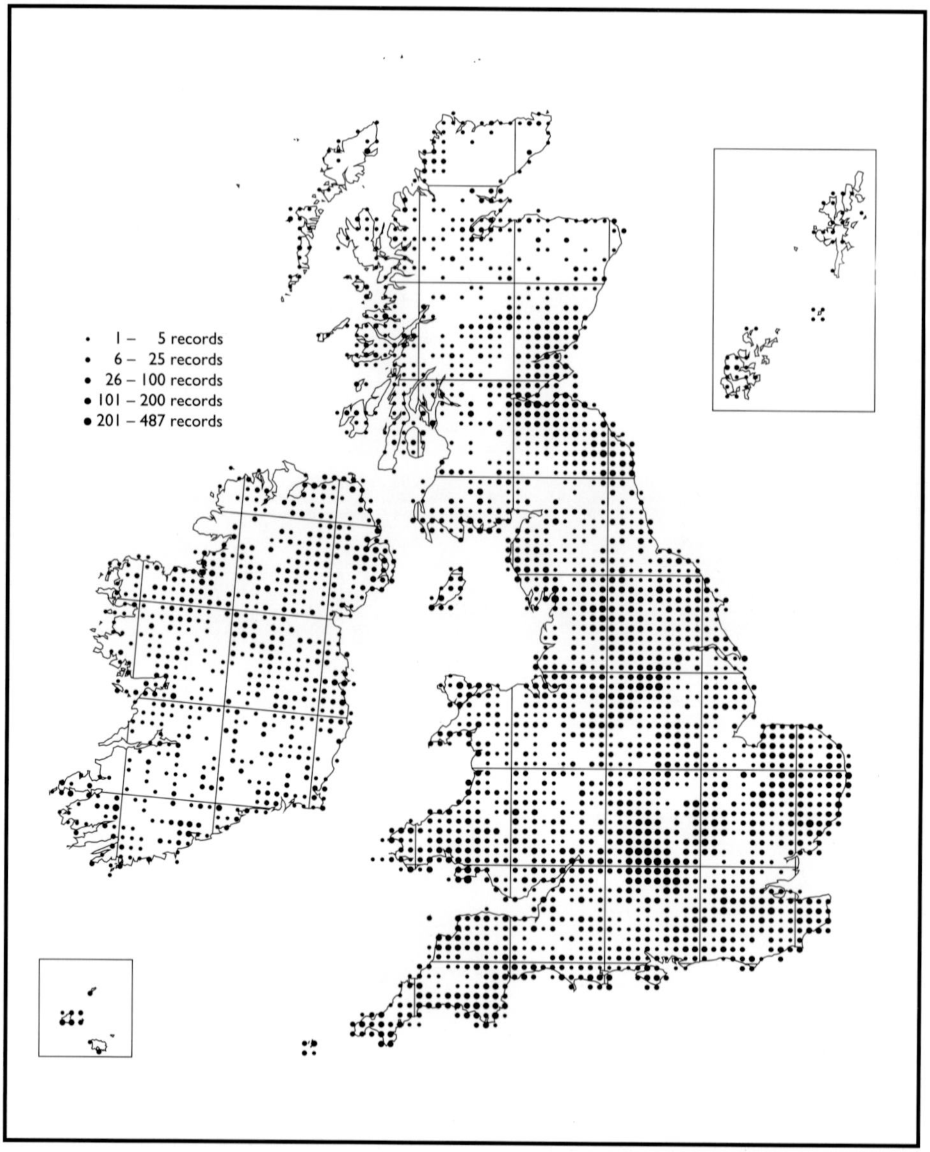

Map 2: Frequency of records map

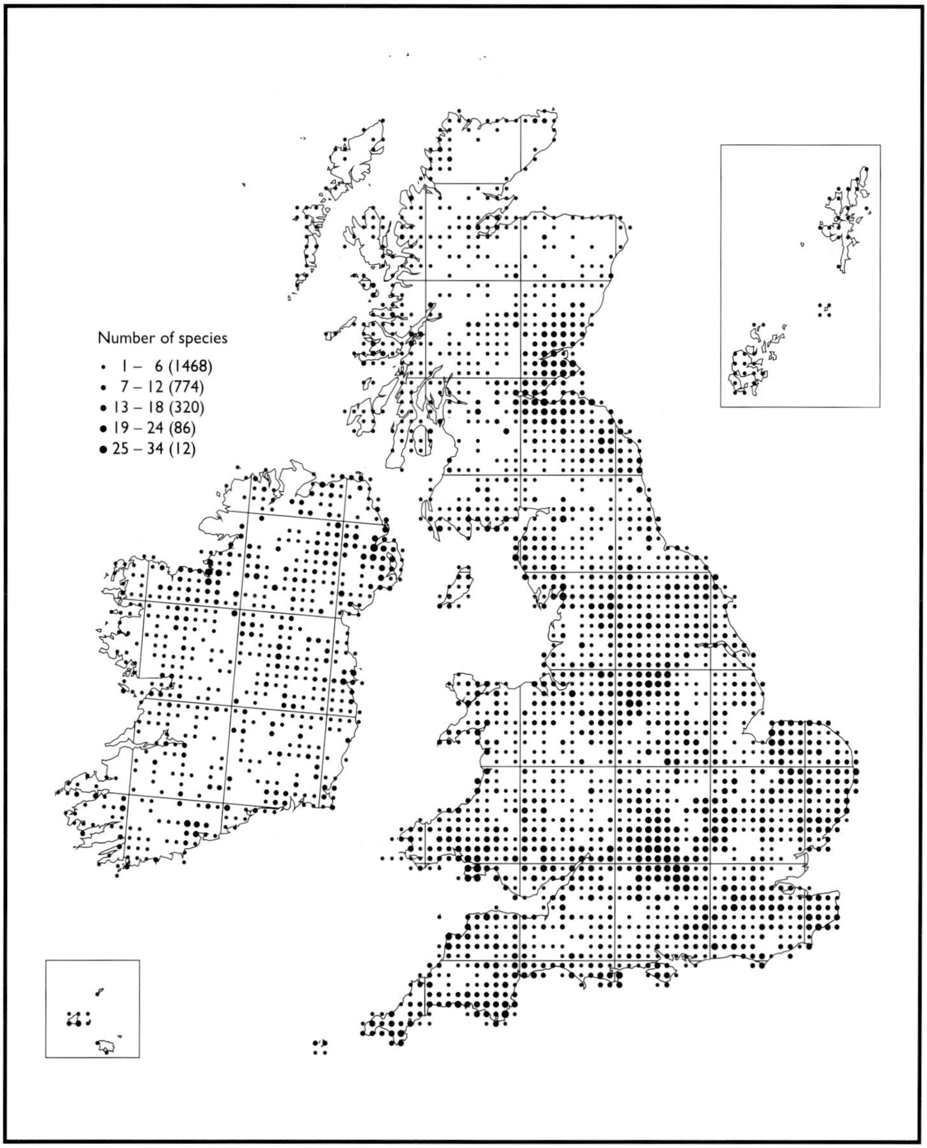

Map 3: Species richness map

Polyxenus lagurus Bristly Millipede

Widespread south of a line from the Severn to the Wash, this species appears to be restricted mainly to coastal sites further to the north and west. Analysis of habitat data suggests the millipede has a very strong association with churchyards and a strong association with buildings. In such situations it can be found by searching at night, especially in humid conditions, when torchlight may reveal aggregations of animals on old stone walls. The habitat data also suggests a strong association with maritime habitats where it occurs under mats of halophile plants, lichens and moss (Blower, 1985). In Belgium the species occurs on old walls but also under bark and lichens on trees (Kime, 2004). Blower (1985) also stated that it was most often found under the bark of dead trees, especially conifers. However, the habitat data suggests a strong negative relationship with woodland. This apparent contradiction is probably explained by the ease of locating the millipede on old walls (relative to the difficulties of searching for it in woodlands) combined with the increased recording effort in churchyards over the lifetime of the recording scheme. Elsewhere the species occurs throughout Europe and also in the Central Sahara and the USA (Enghoff, 1976a; 1978). A parthenogenetic and a sexual form of this species are reported from mainland Europe (Schömann, 1956). The British populations that have been examined appear to be of the sexual form (Fussey & Vardnell, 1980). Adults have been recorded between May and September.

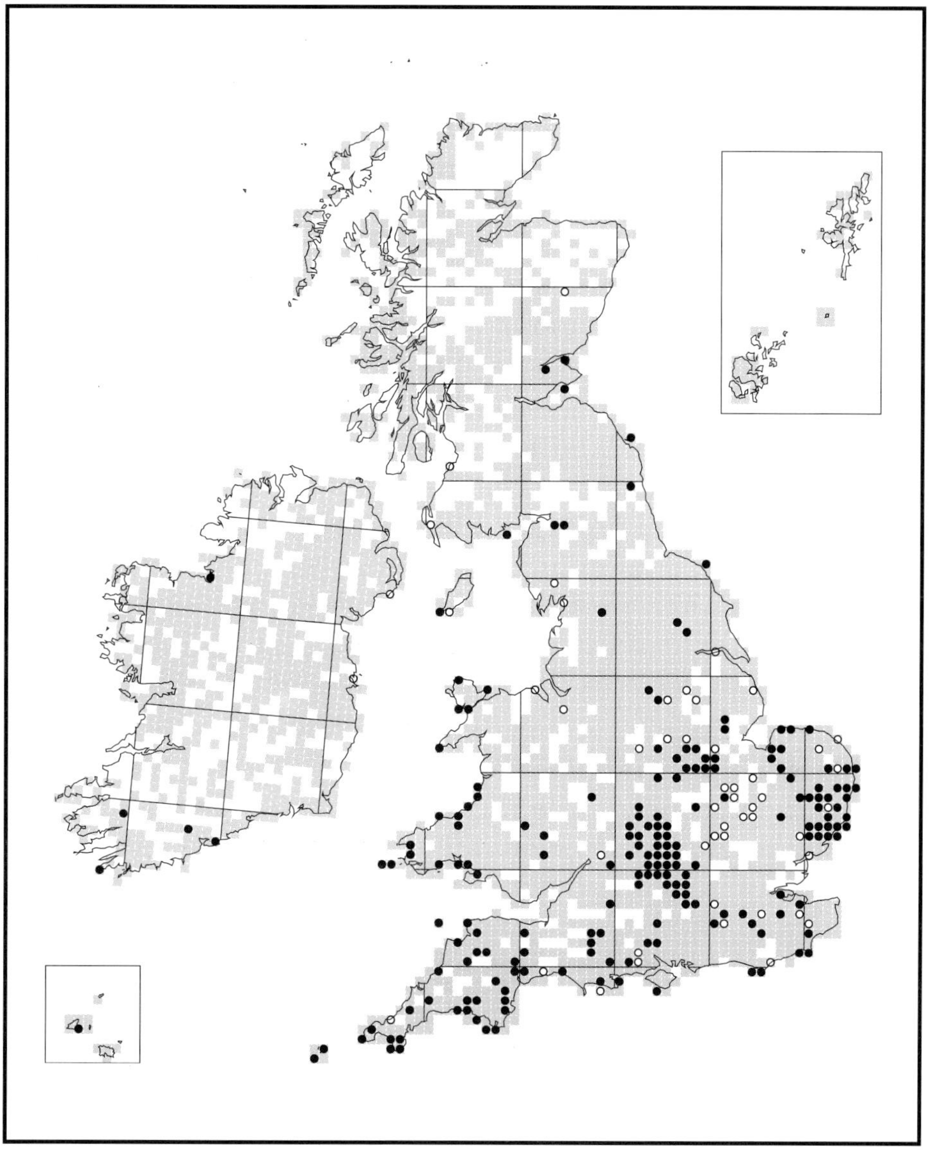

Glomeris marginata **Pill Millipede**

This species is widespread and common in England, Ireland and Wales. Its range extends into southern Scotland but it is not found north of the Midland Valley. An intolerance of prolonged low temperatures probably limits its distribution in the north (Kime, 1995). For the same reason, there are very few British and Irish records of the species from altitudes above 300 m and just one from above 500m. Kime (2004) reported a similar situation from the Ardennes in Belgium. Although it has been recorded from a wide range of habitats and soils, this species is most typically found in deciduous woodland on calcareous loam soils. Habitat data analysis also suggests a strong association with bare rock, including screes, but strong negative relationships with sand dunes and with disturbed, human influenced sites including cultivated land, waste ground and churchyards. In Flanders, Van Den Haute (1999) has shown calcium content of the soil to be the most important factor determining abundance of the species. In Europe this millipede is found eastwards to Poland and from the Pyrenees to southern Norway (Kime, 1990a). It is most common in the west and is restricted to lower altitudes and coastal areas further east (Kime, 1995). Males mature in their second or third year, females in their third or fourth year. Adults can be found throughout the year but are collected most frequently in the spring and autumn. Individuals may survive to an age of 10-11 years (Heath, Bocock & Mountford, 1974).

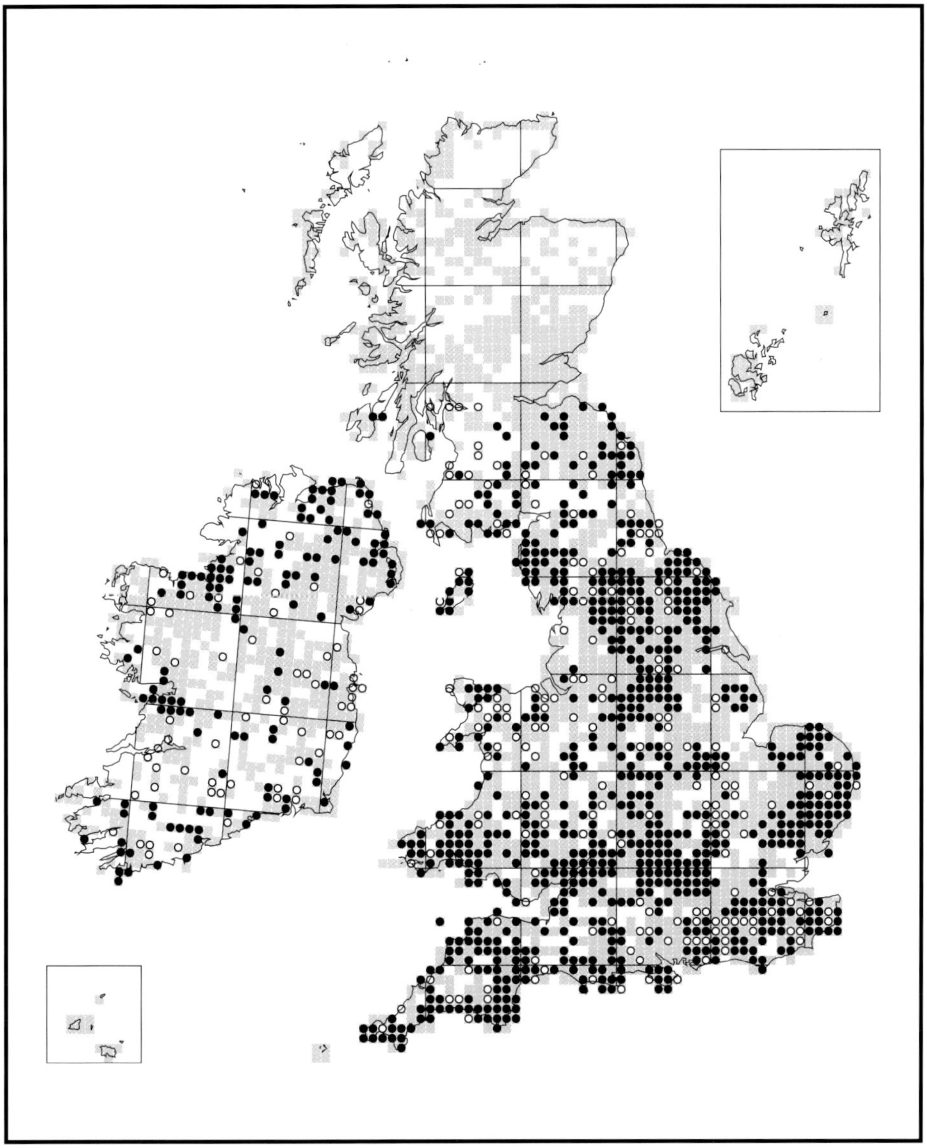

Geoglomeris subterranea
(= *Stygioglomeris crinata*)

This is a soil-dwelling species that has been collected from widely scattered locations across Britain and Ireland. Kime (2004) describes the species as a strict calcicole and 96% of the British records are from calcareous soils. It appears to be absent from large areas of south west England, north and west Scotland and Wales where the geology produces mainly acidic soils. Although it can be common in limestone areas, it has not been found in many apparently suitable sites. A strong association with calcareous soils is not evident from analysis of environmental data due to the small number of records for the species. Analysis of habitat data suggests the millipede has a strong association with excavation sites, especially old chalk and limestone quarries. In Europe there are records from Austria, Belgium, France, (Kime, 2004) Germany and Switzerland (Kime, 2001). Bocock, Heath and Blower (1973) found that this millipede lives mainly in the top 10-20 centimetres of soil, where population densities of up to 128 per square metre were recorded, but may migrate deeper or to other sheltered locations in exceptionally dry or cold conditions. It is a difficult species to locate by hand searching. Branquart and Gaspar (1996) have shown that Berlese extraction is the only method of sampling populations successfully. Most individuals appear to mature in their third year (Blower, 1985) and adults can be found throughout the year. As no male specimens have been collected it is assumed that the British populations of *G. subterranea* are parthenogenetic.

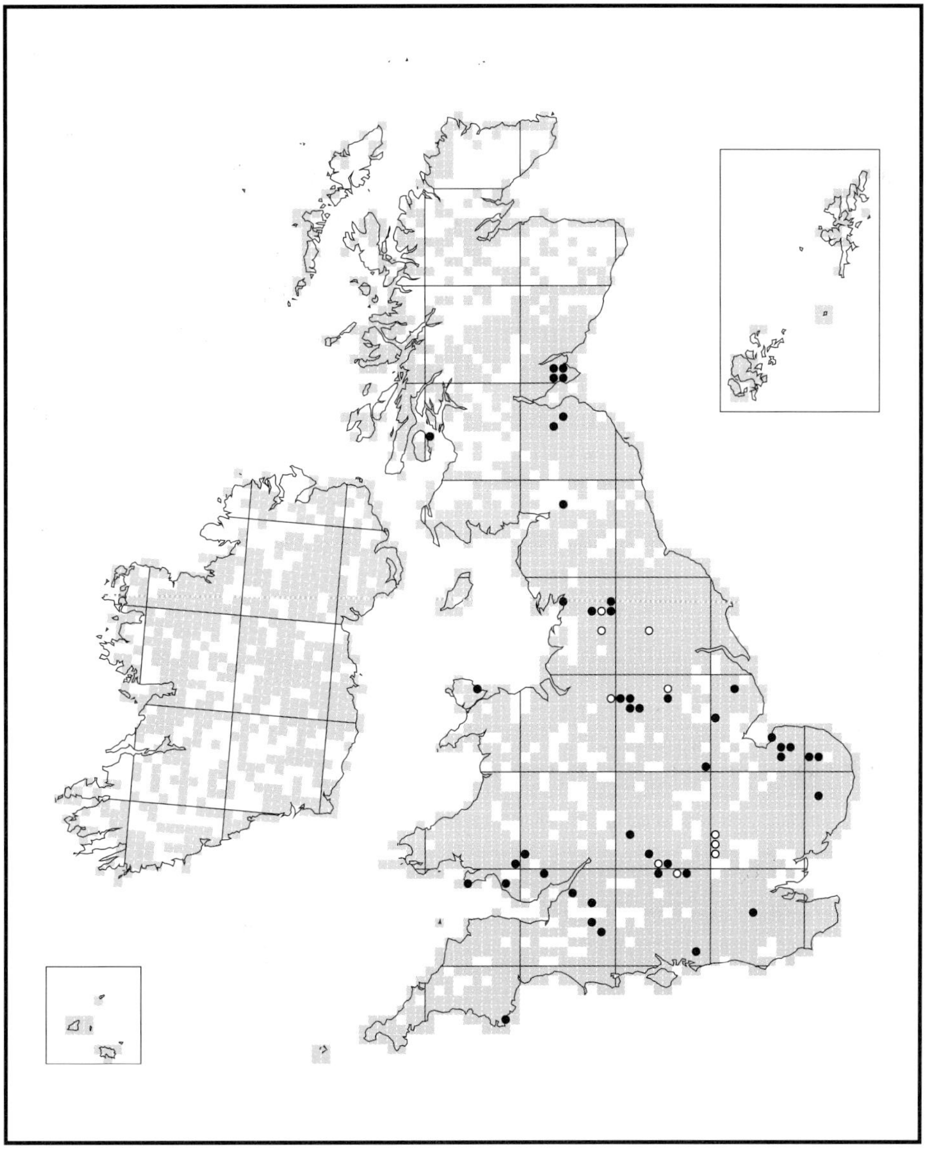

Adenomeris gibbosa

This species was first collected from a site near Dublin in 1978. Further specimens were found on this site in 1979 and another location near Dublin in 1981. During the BMIG Field Meeting in 2004 it was collected with *Geoglomeris subterranea* from Aston Clinton, Buckinghamshire. Further specimens were collected from a nearby site later the same year and from Great Missenden, Buckinghamshire in February 2006. Insufficient habitat data were available for analysis but the British and Irish sites share similarities with the French sites in the western Pyrenees (Mauriès, 1960; Demange, 1981). The French and British specimens were all collected from scrub or woodland on calcareous soils. At least one of the Irish sites was an area of cleared scrub on limestone (Blower, 1985). Although all of the specimens found so far were collected by hand searching (in soil, under stones, under moss, and in leaf litter), it is possible that Berlese extraction or related techniques would be more productive for sampling this species. Outside of Britain and Ireland it is known only from the western Pyrenees. However, Kime (2001) considers it may have been overlooked in western France and may possibly have an Atlantic distribution. The Irish records have included adult females in September, November and February and adult males in February. Adults of both sexes were found in April in Buckinghamshire.

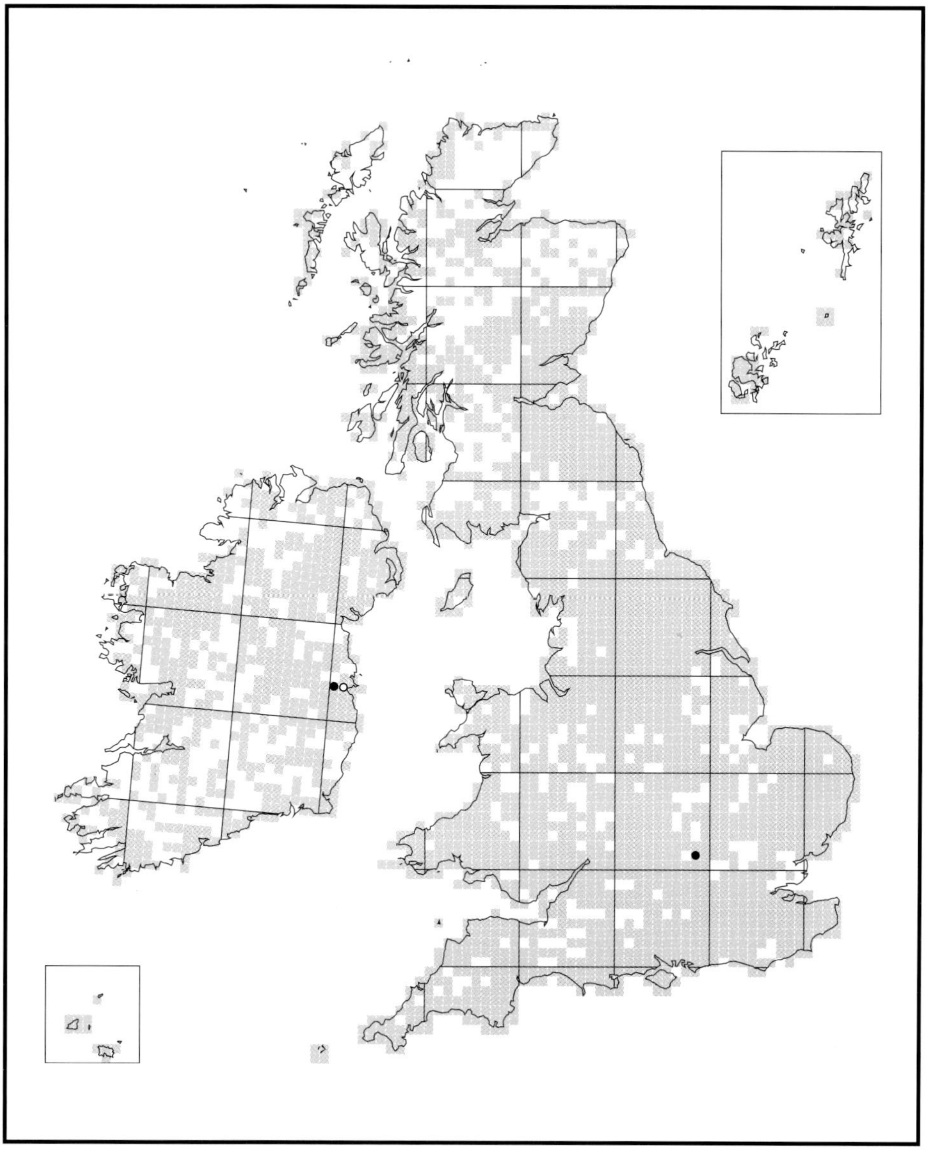

Trachysphaera lobata

The only British records of this species are from two sites near Bembridge on the Isle of Wight. Searches of other, apparently suitable sites have failed to discover the species and one of the original sites has since been lost to development. The remaining site comprises a narrow belt of sycamore woodland at the base of soft earth cliffs. The species is found within a few metres of the high water mark in fallen dead wood, leaf litter and soil. Although the soil is mainly clay the millipedes are only found within pockets of sandy, humus rich soil. Jones and Keay (1986) found the species was most abundant at depths of c. 15 cm in these sand pockets where densities were equivalent to 5 600 m^{-2}. Elsewhere in Europe it is known only from a small number of calcareous sites in western France, mainly in caves, and from woodlands and a quarry near Paris (Kime, 2001). Although it is unknown in the French populations, Bratton (1991) suggested that the British populations of *T. lobata* may be parthenogenetic. However, a male to female sex ratio of 1:3 was found for a sample of specimens collected from the British population in 2005 (Lee *et al.*, 2005). Further examination of these males also suggests that the British population forms a distinct geographical race of the species.

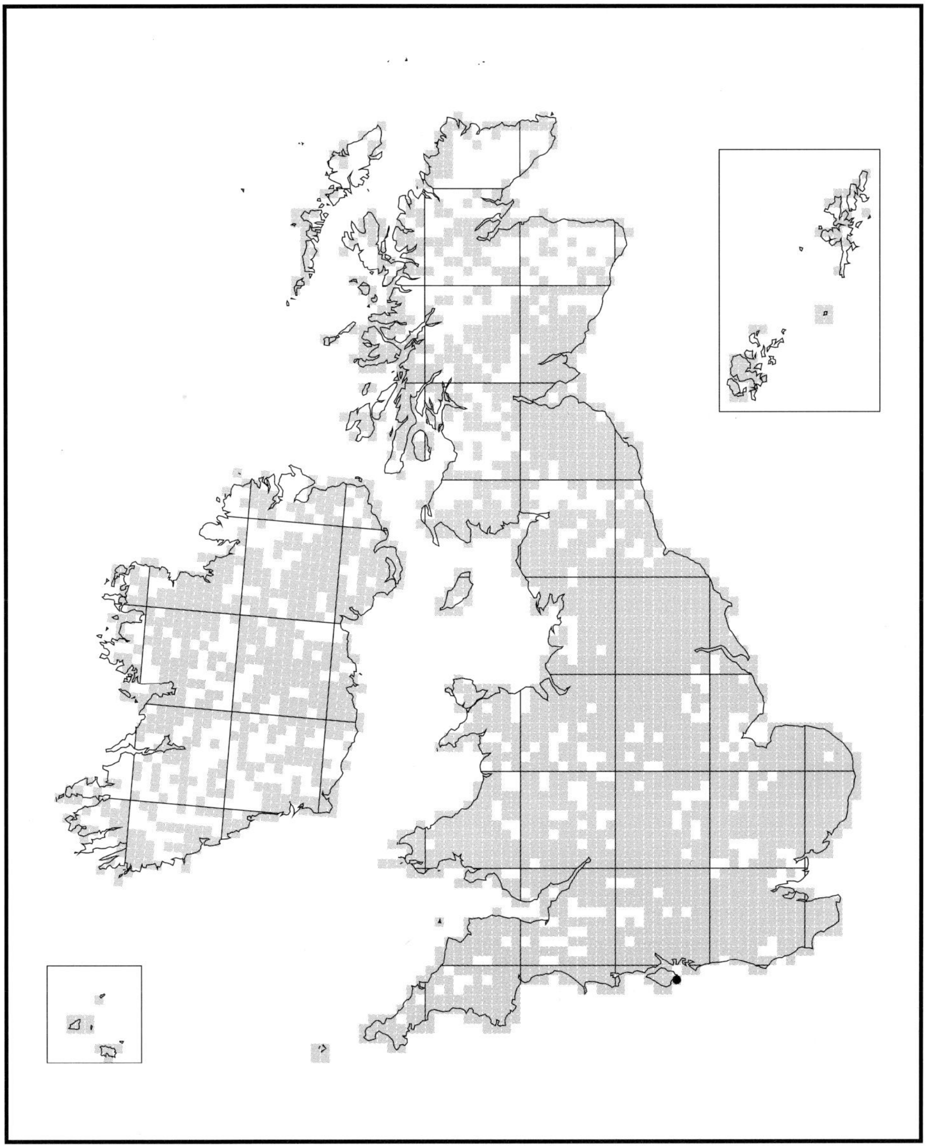

Polyzonium germanicum

There is a single record of this distinctive species from Surrey based on a section of moulted cuticle (Arthur, Cloudsley-Thompson & Sankey, 1951). All other records are from Kent. Even within this restricted area it appears to have declined in abundance and range over the last 25 years. Kime (1990b) argued that it may have been more widespread in Britain 5000 years ago when the post-glacial climate was at its warmest. Insufficient habitat data were available for reliable conclusions to be drawn from the analysis but 88% of the records with habitat data are from woodland. The first British specimens were also collected in mixed deciduous / coniferous woodland on chalk (Brade-Birks, 1920). Most recently the species has been taken by pitfall trapping in stands of chestnut / hazel coppice of different ages. It fell into traps in June, September and October but not July and August. Although present in all ages of coppice, the greatest number of animals occurred in the oldest (15 year old) coppice. There appear to be two main European populations, one in the eastern half of Europe and one in France, just extending into south east England. These populations are separated by a large area of northern central Europe from which there are only a handful of scattered records (Kime, 2001). Schubart (1934) found no differences between individuals from the two populations.

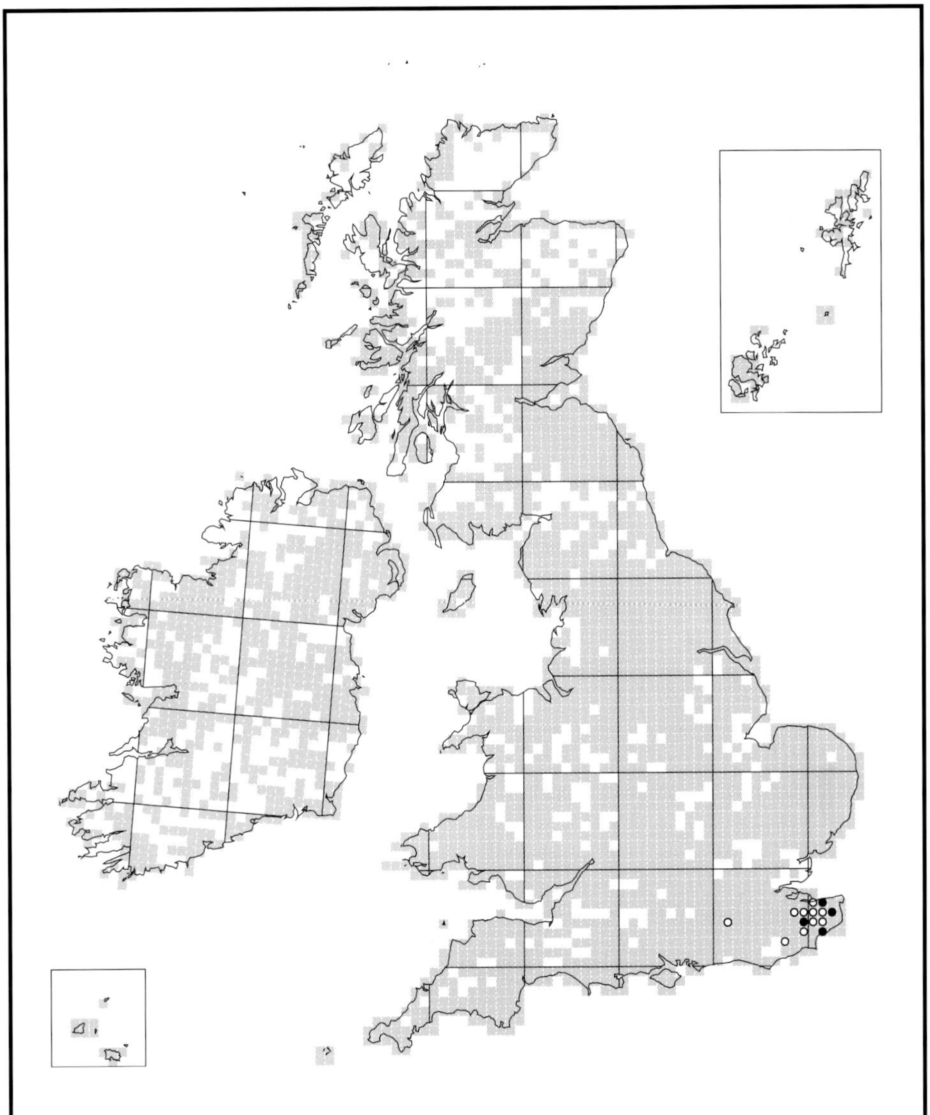

Craspedosoma rawlinsii

This species appears to be widely distributed across Britain and Ireland but is not common. Clusters of records from particular regions largely reflect the activity of a specialist recorder in those areas. Blower (1985) and Kime (2004) considered this species to be primarily a woodland animal but analysis of the habitat data suggests there is a strong association with wetlands. Even in woodland it is associated with microsites which provide the high humidity it prefers (Haacker, 1968). Harding and Jones (1994) noted its presence in East Anglian woodlands subject to considerable summer drought and considered it likely the animal would survive such conditions by burrowing deeply in the light soils. In Belgium, Kime *et al.* (1992) found the species was associated with soils of high sand and low calcium content. It is also reported from sites with acidic, sandy soils in Switzerland (Pedroli-Christen, 1977) and Germany (Spelda, 1999). Although 80% of British and Irish records for which data were supplied were from non-calcareous soils, a strong association is not evident from the analysis of environmental data due to the small number of such records. *C. rawlinsii* (including the subspecies *alemannicum* (Verhoeff, 1910)) has a wide distribution in central Europe but it does not extend into western France (Kime, 2004). Both subspecies are present in Belgium but all British and Irish animals examined have been of the subspecies *rawlinsii*. The species has been recorded throughout the year but adults are usually found during the winter months.

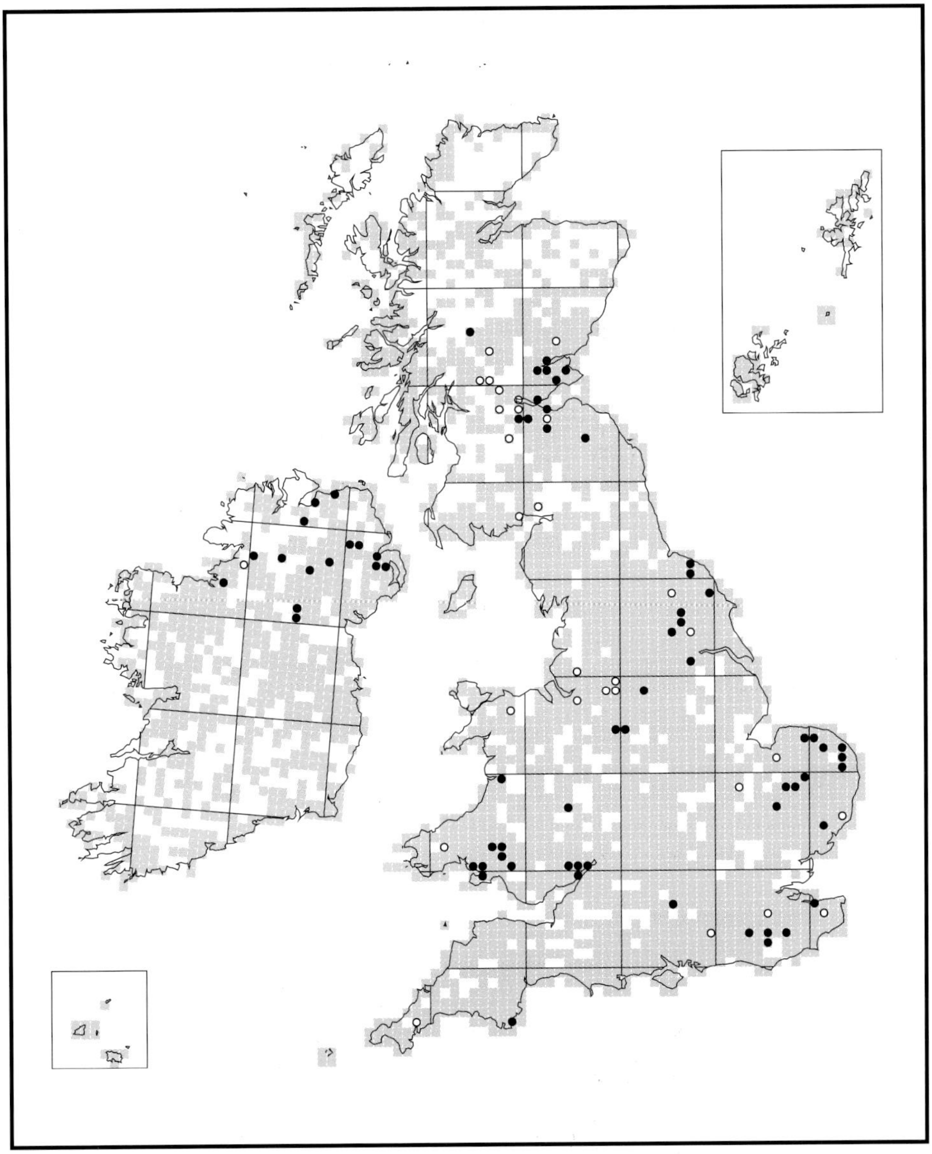

Nanogona polydesmoides

This species is common and widespread throughout Britain and Ireland. Blower (1985) noted it was the commonest millipede recorded from British caves, a habitat more typical in the south of its European range (Kime, 2001). Analysis of the habitat data confirms a strong association with caves and also with waste ground, especially in urban areas. However, it is found in a wide range of other habitats and has usually been considered a species of deciduous woodlands. Analysis of the environmental data shows, at most, a slight preference for calcareous soils but various authors have noted such a preference (Blower, 1985; Gregory & Campbell, 1996; Kime, 1978). Gordon Corbet (pers. comm.) reports the species as rare in non-calcareous natural sites in Scotland but frequent where there is concrete or mortar. Due to the strong association of the millipede with urban waste ground, these latter records may be masking any association with calcareous soils. Kime (2004) describes its continental distribution as strictly Atlantic. Outside of Britain and Ireland it occurs mainly in France but is very local in Belgium, there is a single record from the west coast of Norway and an outlier population apparently belonging to this species is found in northern Italy. Adults have been found throughout the year but are most frequently recorded in the autumn and winter months. Although some individuals may survive for longer, Blower (1985) considered the species to be annual with egg-laying taking place from November to January.

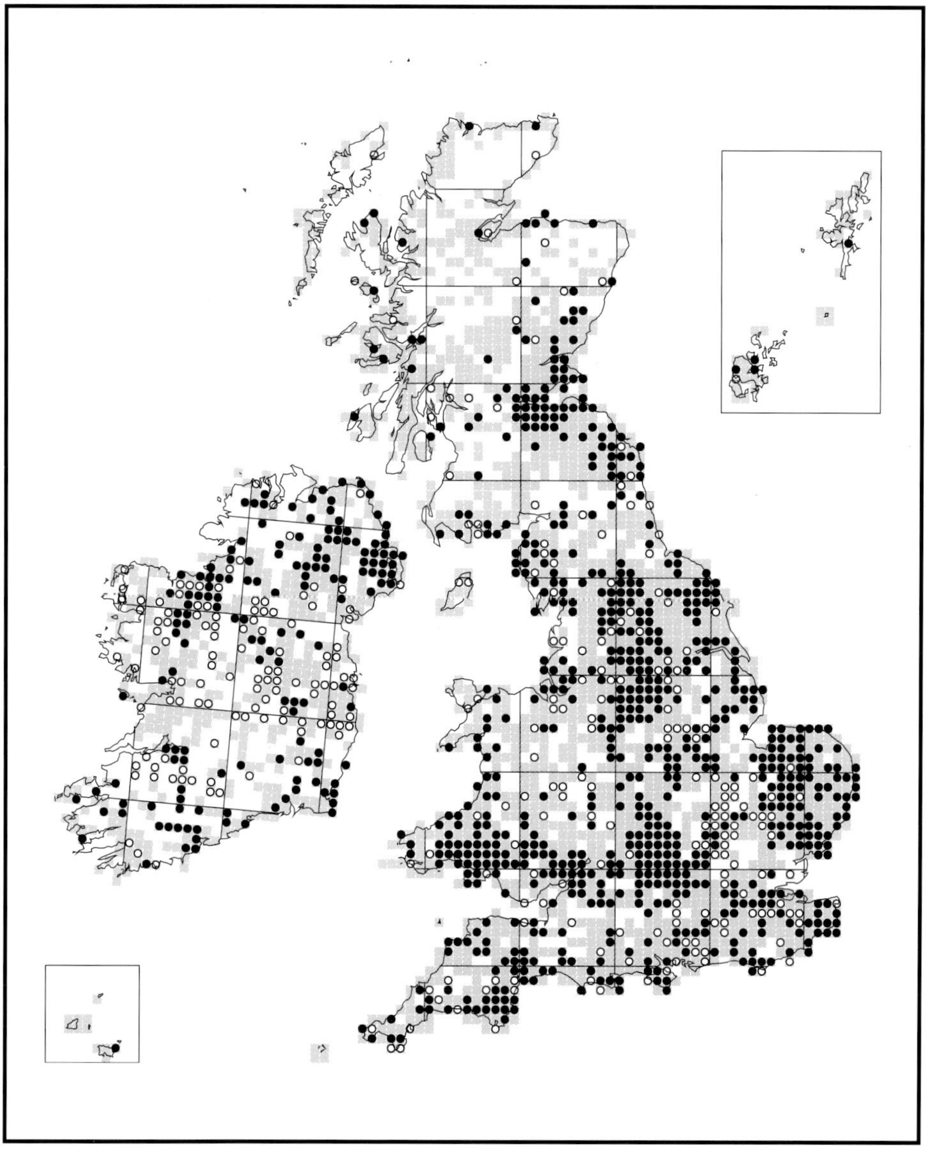

Anthogona britannica

This species was only recently described as new to science from specimens collected at Slapton Ley, Devon. Although first collected in 1983, the specimen had been assumed to be an immature *Craspedosoma rawlinsii* until the collection of further examples in 1992 (Gregory, Jones & Mauriès, 1993). Bolton (1996) gives details of further sites in the Dartmouth area but the species has not been found outside of south Devon (VC 3). It has been collected from a variety of habitats including coastal cliffs, grassland, vegetated shingle and woodland. Insufficient data were available for reliable conclusions to be drawn on habitat associations, indeed Gregory, Jones and Mauriès (1993) concluded the millipede had no strong habitat preferences. However, it is worth noting that all of the records are from coastal locations and the majority of records have been from leaf litter samples below deciduous trees or a ground cover of ivy (Bolton, 1996). The genus *Anthogona* contains just two species and is restricted to Europe where it shows an Atlantic distribution. The second species, *A. variegata* Ribaut, 1913, has been found in France and Spain (Enghoff & Kime, 2005). When first discovered, *A. britannica* was expected to occur on the other side of the English Channel but, so far, all adult specimens of *Anthogona* from France have proved to be *A. variegata* (Kime, 2001). As a result this is the only British millipede that could be considered endemic. Mature adults have been found between October and February. The animal almost certainly has an annual life cycle.

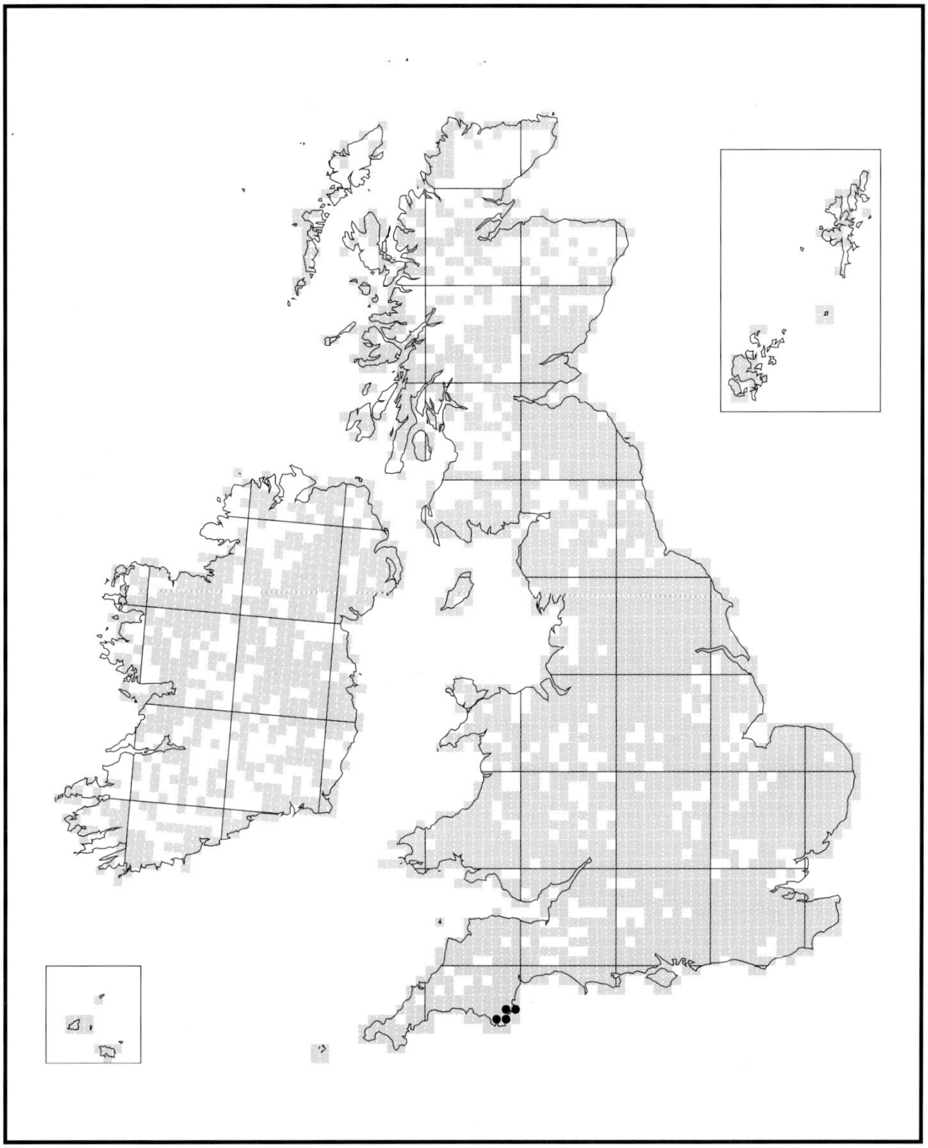

Chordeuma proximum

This species has a distinct south western distribution in Britain with single records from Carlisle and the Isle of Rum, Inner Hebrides. There is evidence that the millipede has been introduced at these northern locations. It may also have been introduced to some, if not all, of the sites in the north of Ireland. Analysis of the habitat data suggests a strong association with woodlands and Kime (2001) describes it as a typical woodland animal in France. A strong association with coastal sites is suggested also but such a link is not apparent in France and may be an artefact caused by the south western distribution in Britain. Kime (1990a) associated this species with acidic sands or loams. Those records providing data on soil factors support a possible link with acidic soils (67% of records) and with loams (55% of records) but there is insufficient evidence to confirm strong associations and there is no evidence of a link with sandy soils. Elsewhere in Europe this species has been found only in France and shows a classic Atlantic distribution. Unlike in Britain, its range in France lies to the west of that of *C. sylvestre* although they overlap in Normandy (Kime, 2001). It is an annual species in the south although Blower (1985) suggests it may have a two year life cycle further north in the Forest of Dean. Adults are found most frequently in the winter (Read, 1988) but have been recorded from September through to June.

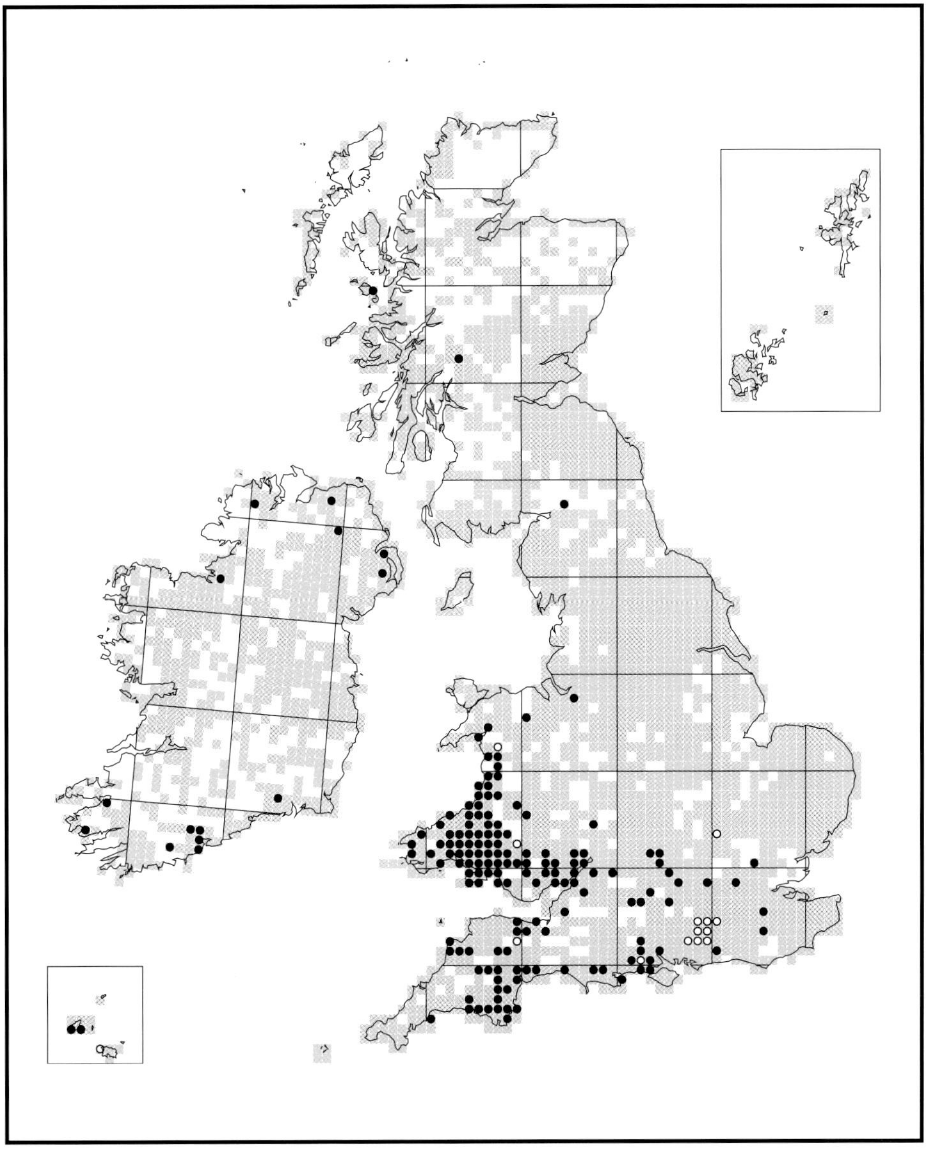

Chordeuma sylvestre

In Britain this species appears to be restricted to Cornwall but it has a much wider European distribution. Its ecological requirements may explain its presence in Cornwall but not its absence from the rest of the south west or from southern England generally. In Cornwall the millipede has been collected from woodland, waste ground and maritime habitats but insufficient habitat data were available for reliable conclusions to be drawn from the analysis. Although a common woodland animal in the Limburg province of the Netherlands (Jeekel, 2001) and in most of Belgium, Kime (2004) did not consider the species to be a habitat specialist other than in respect of an association with rocky ground. He reported a wide range of other habitats from which it had been collected, most of them providing a deep litter layer. The litter is important as it is soil moisture rather than soil acidity or soil type which appears to be a key factor in determining distribution (Kime & Wauthy, 1984). This millipede is essentially a central European species. In the north its range reaches Springe, south of Hanover. The north western edge of its range extends to at least Normandy and hence the Cornish populations may be a natural extension of this range (Kime, 2001). Adults of this animal have been found in April but only sub-adults in autumn suggesting it is an annual species (Blower, 1985).

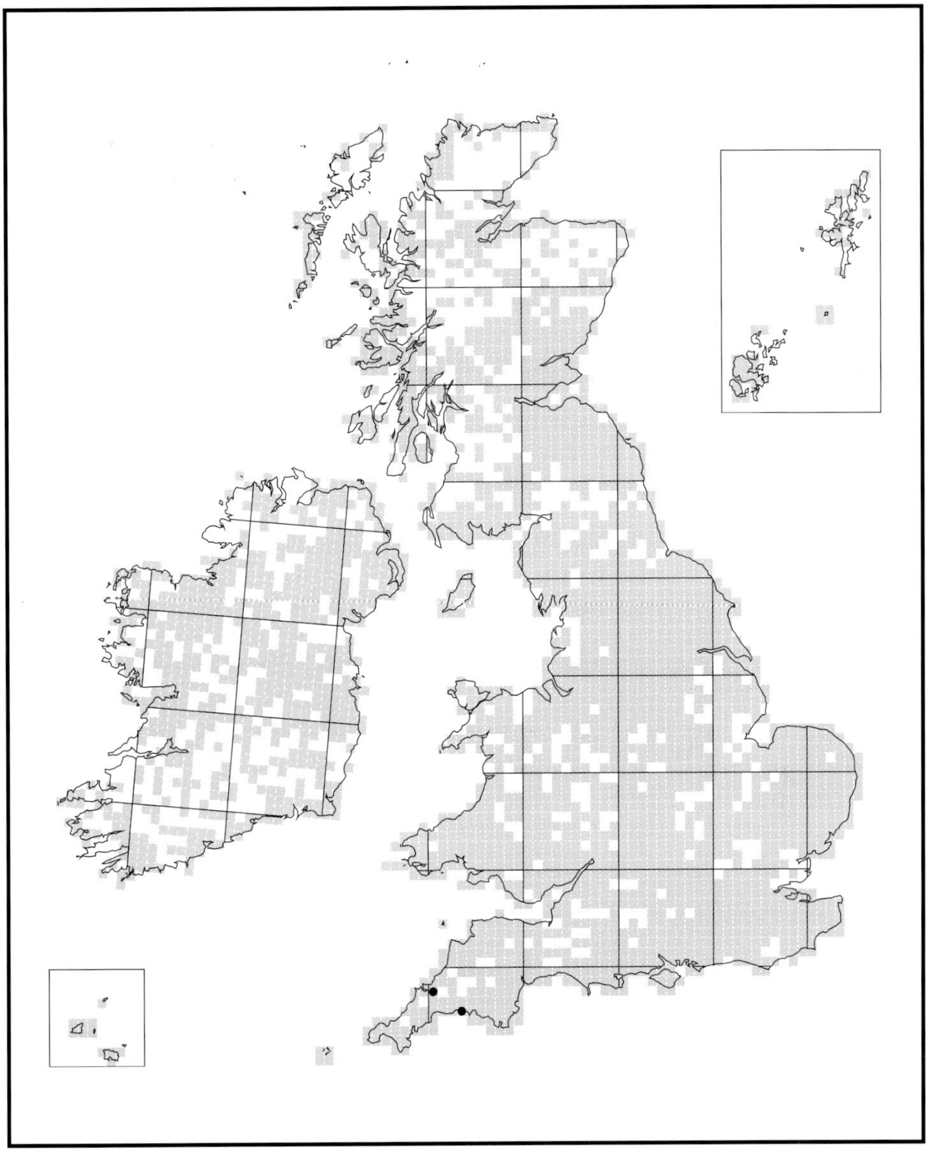

Melogona gallica

The distribution pattern of this species has become somewhat uncertain with the recent discovery of *M. voigti* in Britain. Although the map shows all records submitted under the name *M. gallica*, most of these have not been confirmed by examination of male specimens and may include records of *M. voigti. Melogona gallica s.l.* is most frequent in the south and west of Britain but surprisingly there are records from just two sites in Ireland. Its absence from East Anglia and Kent may be due to the drier climate compared with western areas (Kime, 1995). Kime (1990a) considered it a common woodland species and, although 64% of British records were from woodland, analysis of the habitat data suggests no strong relationships with any habitat. In Belgium the species has been linked with lower temperatures (Kime & Wauthy, 1984). In Europe it has an extended Atlantic distribution occurring from southern France north to Netherlands and east to the Rhine valley (Kime, 2001). A record from Norway probably results from an introduction (Meidell, 1968). It has an annual life cycle (David, 1984), at least in the south of its range, but Meyer found evidence suggestive of a two year cycle in samples from Delamere Forest, Cheshire (Blower, 1985). Adults have been found from October through to June but are most frequent in the winter and early spring.

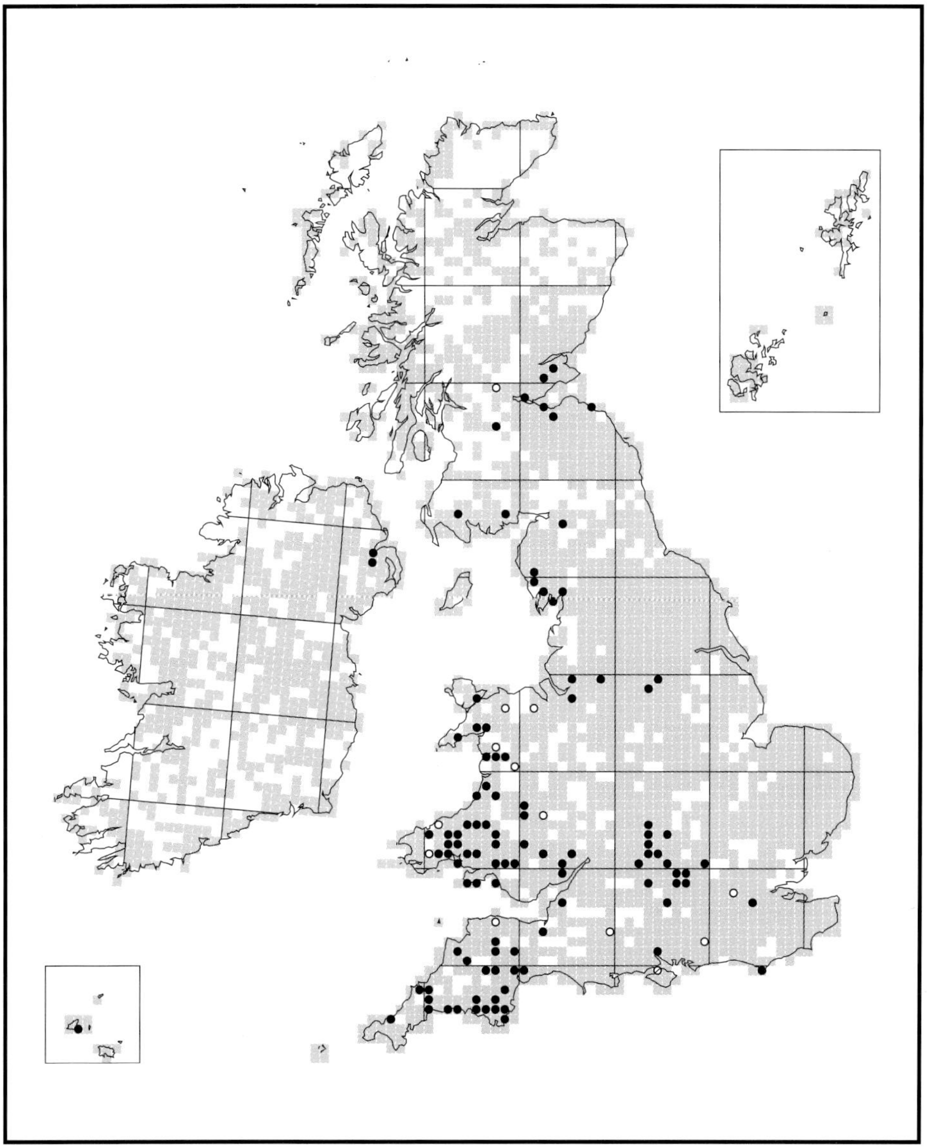

Melogona scutellaris
(= *Melogona scutellare*)

This is a widespread species in Britain and Ireland but it appears to be absent from East Anglia and is generally more common in the north of England than the south. It also seems to be absent from much of the north of Scotland despite considerable effort to find the animal there (Gordon Corbet, pers. comm.). Although typically found in woodlands (56% of records), it also occurs in more open and in disturbed habitats including gardens. Analysis of the habitat data shows no strong relationships with any habitat type. Nor does the analysis suggest a preference for a particular soil type or acidity but there is a surprisingly strong association with suburban sites. The European distribution of the species is discontinuous. It appears to be most common in Britain and Ireland with a few records from the Alps and the Pyrenees (Kime, 2001) and a single one from Belgium (Van Den Haute, 1999). Blower (1979) found the species has an annual life cycle with adults appearing by October in south Wales and slightly later in the year further north in Derbyshire. He recorded population densities of up to 64 per square metre in south Wales and up to 109 per square metre in Derbyshire. Adults have been found from September through to May but are most frequent in the winter and early spring.

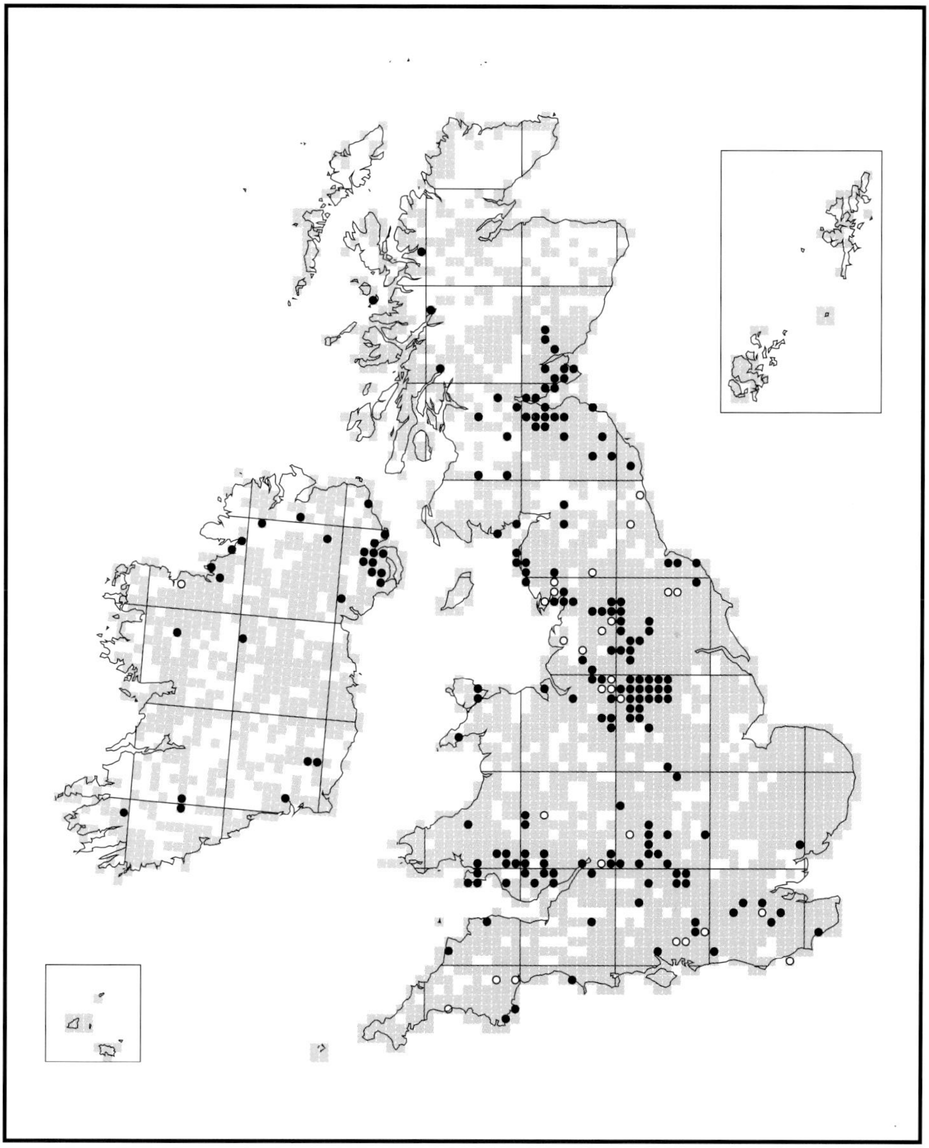

Melogona voigti

This species was not recognised from Britain until 1995 although the earliest specimens known were collected in 1988 (Corbet, 1996). At present it is known from several sites in the Lothians but some of the records of *M. gallica s.l.* may well refer to *M. voigti,* especially if they were based on adult female specimens only. Insufficient habitat data were available for reliable conclusions to be drawn from the analysis but Spelda (1999) considered it to be a woodland species, particularly in calcareous areas. Jeekel (2001) described it as synanthropic in the Netherlands and believed it had been spread to Scotland by the horticultural trade. The Scottish records are from an urban garden and from rural deciduous woodlands with evidence of human disturbance in the form of fly-tipping (Corbet, 1996). *Melogona voigti* is a Central European species which occurs to the north and east of *M. gallica* in Europe (Kime, 2004). Their ranges overlap in Belgium, Netherlands and Scotland and Kime (2001) believed the occurrence of *M. voigti* was 'entirely logical' even suggesting that populations may exist in Eastern England in addition to the known Scottish locations. Adult animals have been found in January, April and May (Corbet, 1996). It is likely the species has an annual life cycle as is the case with our other members of the genus.

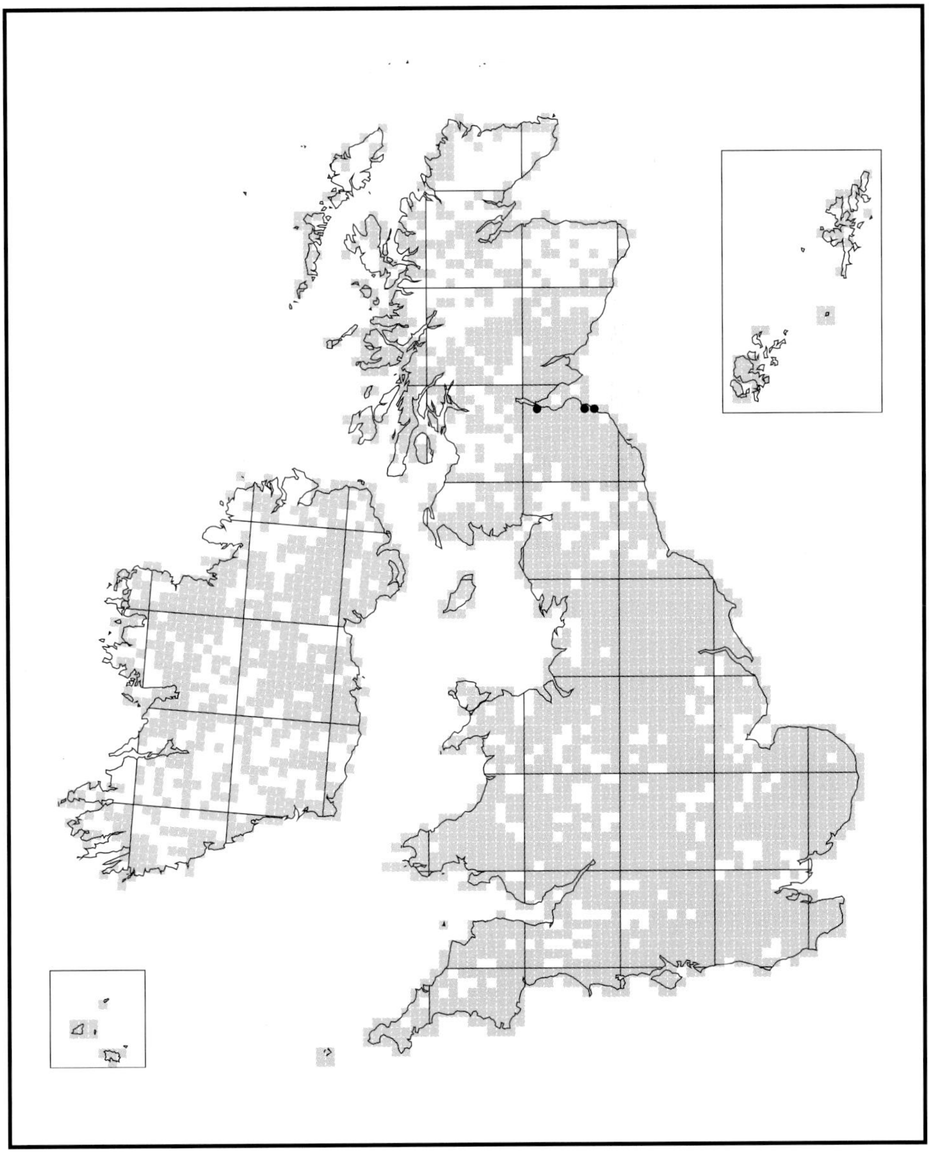

Anamastigona pulchella
(= *Anamastigona pulchellum*)

The first Irish records of this introduced species were from County Down and County Antrim in 1994. It now appears to be well established as a naturalised component of the fauna of at least ten sites in Northern Ireland. Most recently it was found in Londonderry but to date it has not been found in Britain. Insufficient habitat data were available for reliable conclusions to be drawn from the analysis but most of the Irish records are from deciduous woodland and Anderson (1996) suggests that deep, stable litter is preferred. The genus in general occurs in woodland but there are few references to the specific habitat of *A. pulchella* in its native Italy. The few that exist suggest that the Irish sites are typical for the millipede and it is most probably a species of the woodland litter. There is also one record from Mediterranean maquis vegetation (Desmond Kime, pers. comm.) suggesting it may be able to colonise areas of drier scrub in Ireland. The species has been introduced to southern France (Mauriès, 1960) and Maderia (Demange, 1970) as well as Ireland. In Ireland adults have been collected from October to February but there is some evidence of a longer than annual life cycle (Anderson, 1996).

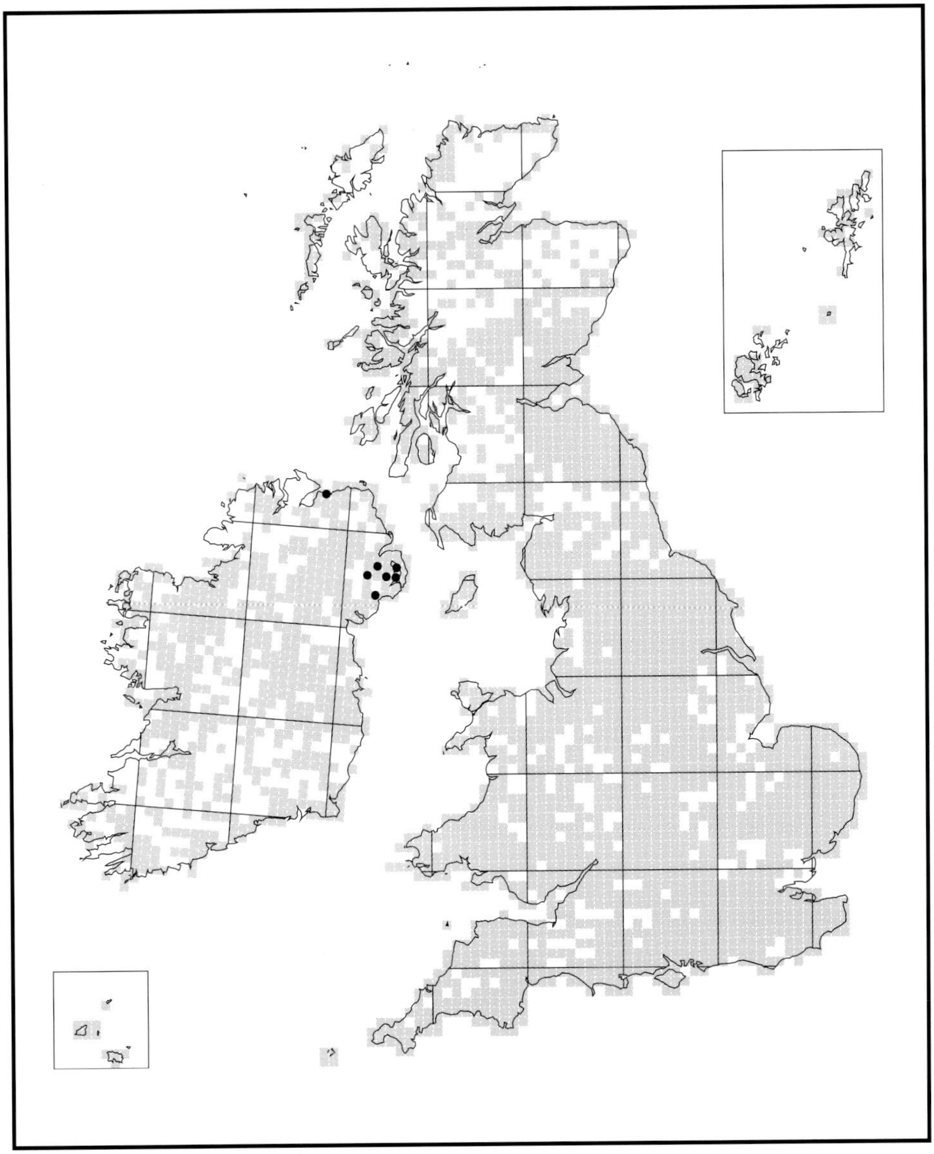

Brachychaeteuma bagnalli

Although not common, this species appears to be widespread in Ireland, northern England and southern Scotland. Its range overlaps with that of *B. bradeae* in north west England and southern Scotland. This more westerly distribution of *B. bagnalli* in Britain and Ireland fits a distribution pattern recognised by Kime (2001) as being 'marginal between Atlantic and Central' in Europe. However, the two taxa have very variable gonopods and may yet prove to belong to a single, variable species (Blower, 1986). Analysis of the habitat data suggests that this is a garden species as it shows a strong association with cultivated sites in urban and suburban locations. In Britain and Ireland it appears to live in the litter and upper soil layers in these synanthropic sites and it has been found in similar situations in the suburbs of Brussels. Elsewhere in Europe this species has been collected from caves in Westphalia, Germany (Schubart, 1938) and in limestone areas of the Mosan District of Belgium (Kime, 2004). An animal discovered near Lyons, France probably also originated underground as it was found during filtration of spring water (Geoffroy, 1992). Kime (2004) noted that the millipede occurs in deep galleries in these caves, not just at the entrance, and it is apparently well-adapted for its cavernicolous existence. Most observations of adults of this annual species occurred in winter and spring but individuals have been recorded in June. Even in sub-zero temperatures, it can be collected from the underside of surface debris such as logs and stones.

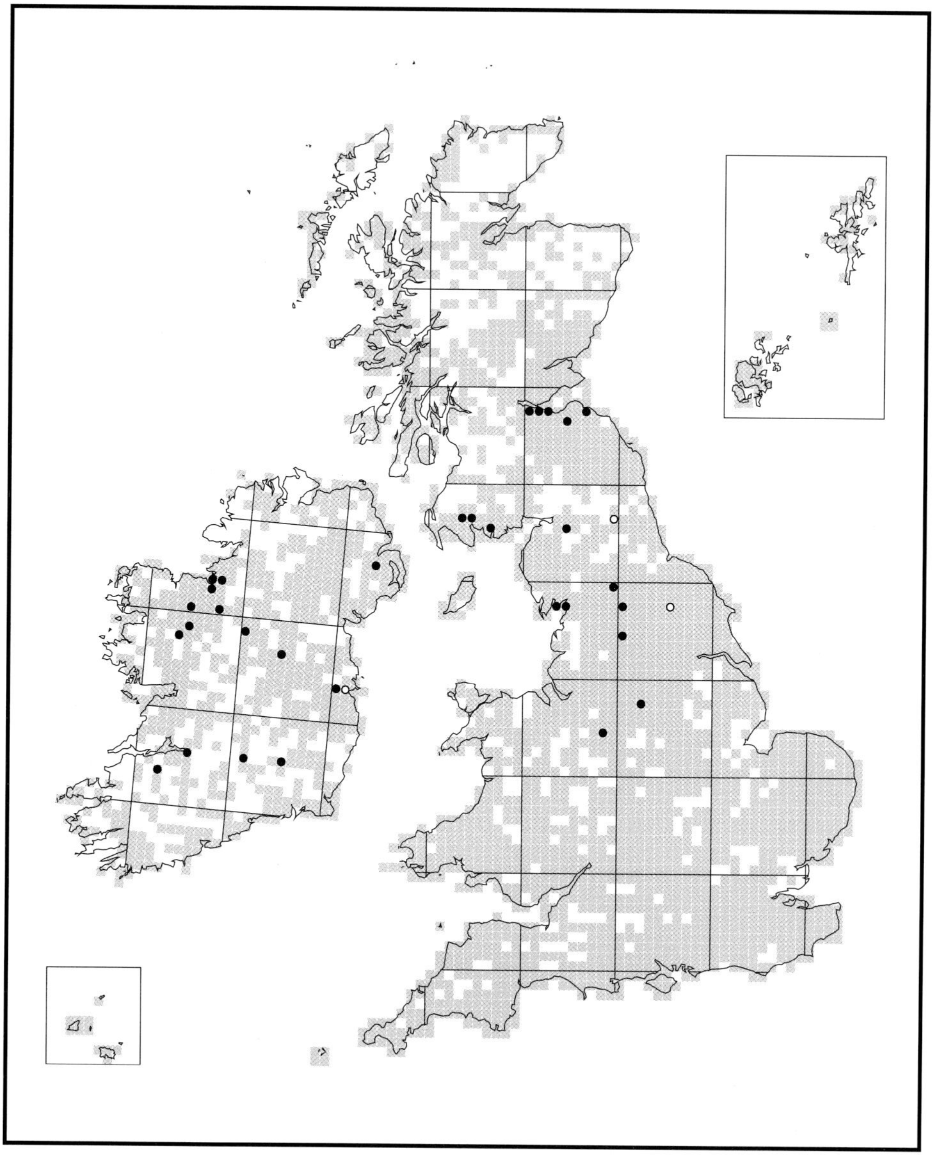

Brachychaeteuma bradeae

Although its range overlaps with that of *B. bagnalli* extending northwards to Perthshire, this species has been recorded most frequently in the Home Counties and East Anglia. It is apparently absent from Ireland. The more easterly distribution of *B. bradeae* in Britain fits with the Central European distribution pattern recognised by Kime (2001). However, the two taxa have very variable gonopods and may yet prove to belong to a single, variable species (Blower, 1986). Gregory & Campbell (1996) typically found *B. bradeae* in ancient woodlands on clay in Oxfordshire but analysis of the habitat data suggests a strong association with cultivated sites. The strong synanthropic association shown by *B. bagnalli* is not so evident for *B. bradeae* but this may be due to the smaller number of records. Many of the records from continental Europe are from synanthropic situations (Kime, 2001) but Schubart & Husson (1937) found this species in caves in north west France. It also occurs in Germany (Schubart, 1930), the Netherlands (Jeekel, 2001) and Sweden (Lohmander, 1925). As with *B. bagnalli,* this is an annual species that is adult in the winter and early spring although individuals have been collected as late as June. It is usually collected from the underside of surface debris such as logs and stones but Jeekel (2001) found it in moles' nests on peaty soil in meadows near Amsterdam.

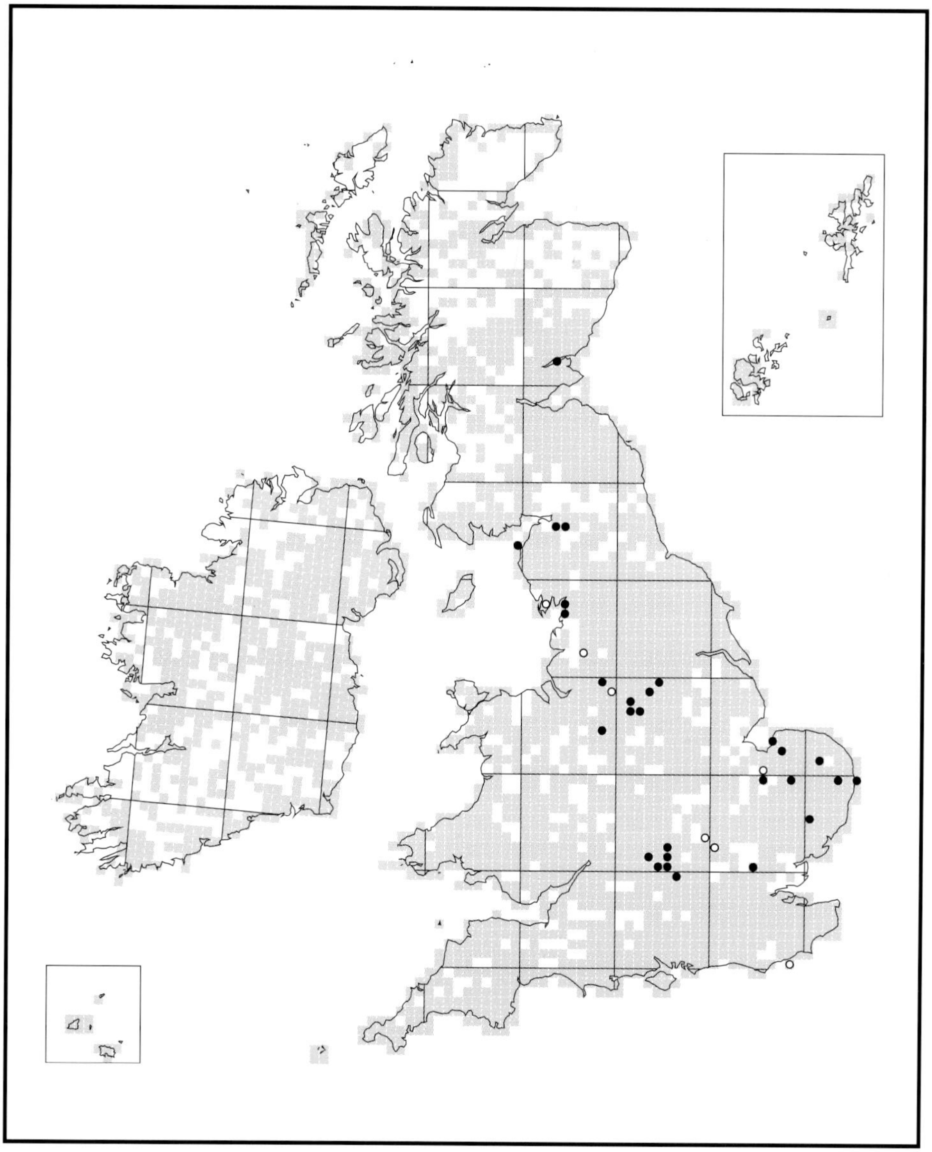

Brachychaeteuma melanops

This species does not appear to be common but it has been collected from localities spread across the south of England and south Wales. It has also been found in the south of Ireland. Its distribution has a more distinct southern bias than the other two members of the genus although its range overlaps with that of *B. bradeae*, especially in East Anglia. This pattern accords with a strict Atlantic distribution in Europe as noted by Kime (2004). Analysis of the habitat data suggests that this is a garden species as it shows a strong association with cultivated sites in urban and suburban locations. Kime (2001) reported it as common in woodland on calcareous soils in northern Aquitaine, France. In Britain and Ireland also, it is recorded most frequently from calcareous soils (76% of records with soil acidity data) but a strong association is not evident from the analysis of environmental data due to the small number of such records. Elsewhere in Europe the species appears to be restricted to the milder parts of western France (Kime, 1995). As with other species in the genus, this animal has an annual life cycle. Adults have been collected in each month from October through to May but most frequently in March and April.

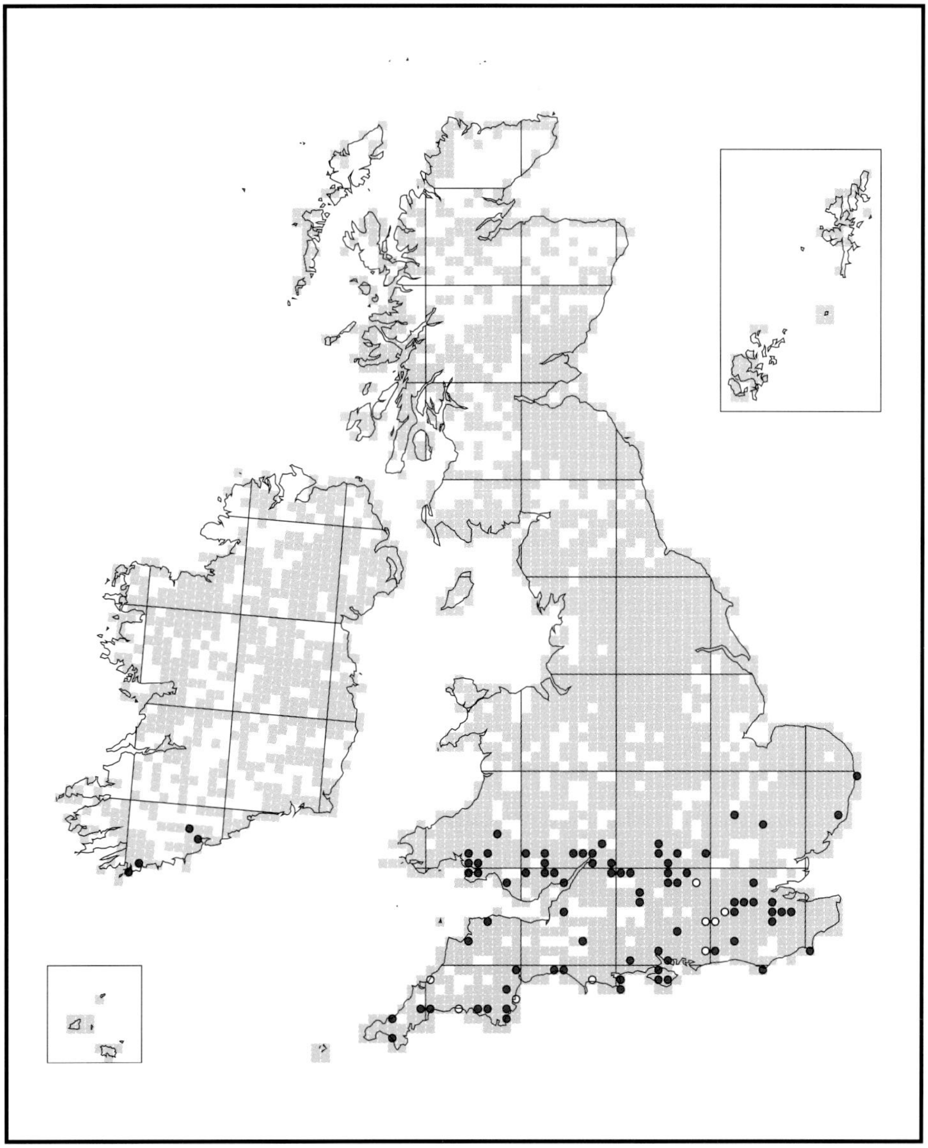

Stosatea italica

This species is presumed to be an ancient introduction. Although it has been most frequently recorded from Kent, there are scattered records from south east England, from around the Severn estuary and from southern Ireland. There is also an isolated record from the Shrewsbury area. Kime (1995) noted that the species appears not to persist at a site for long but it has been recorded from at least one of the sites in Kent over a period of many years. Although Kime (1978) and Blower (1985) associated the species with synanthropic sites, only 35% of records submitted to the recording scheme were from urban or suburban locations. However, 92% of the records were from calcareous soils, a finding that is in agreement with Kime (1978). There are insufficient habitat data available for reliable conclusions to be drawn from any further analysis. The natural range of the species is in Italy and neighbouring areas of the Mediterranean (Kime, 2001) but it has been introduced to synanthropic sites in countries further north including France (Demange, 1981), Luxembourg (Remy & Hoffman, 1959) and Switzerland (Pedroli-Christen, 1993). It has also been introduced to Madeira (Blower, 1985). In Britain and Ireland adults have been recorded in January, April, May, July and October.

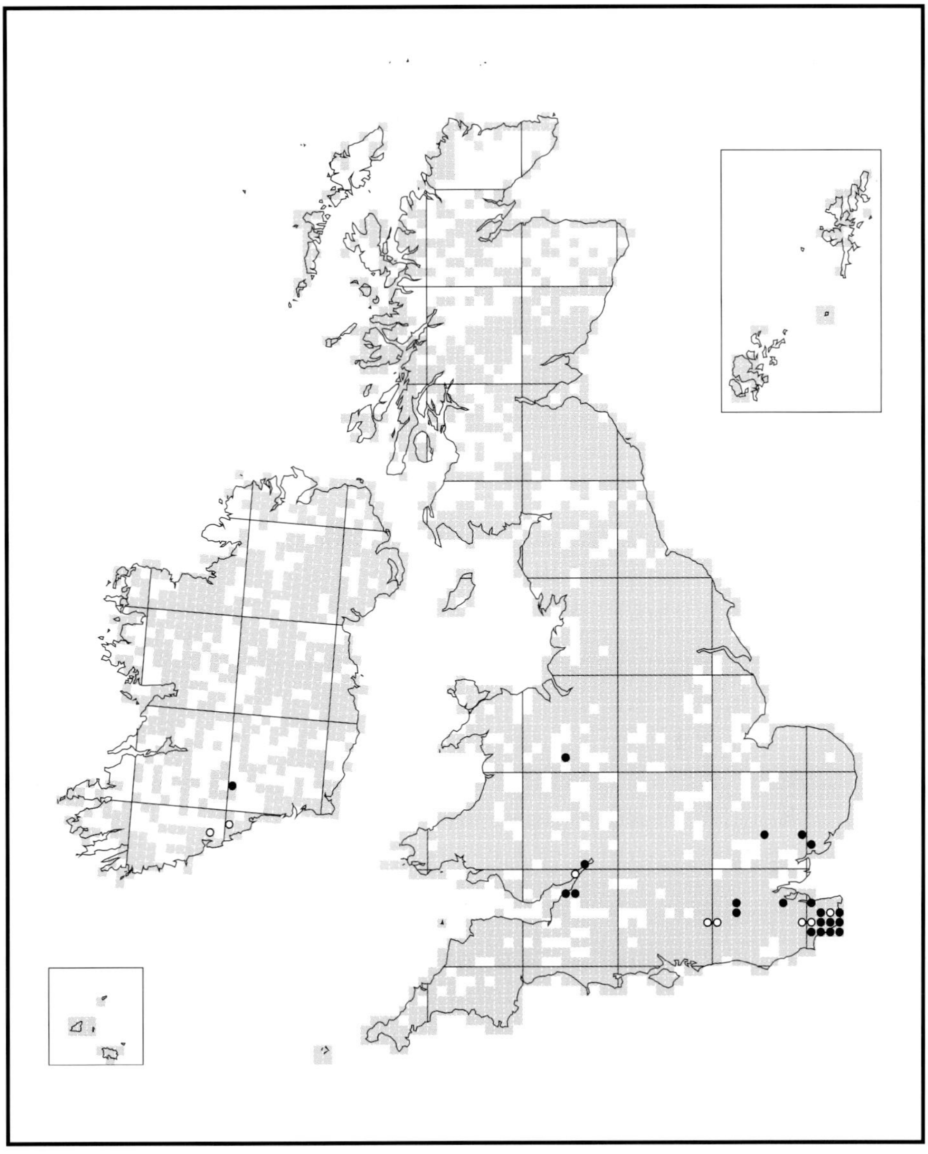

Brachydesmus superus

This is a common and widespread species in Britain and Ireland. It is a generalist which has been recorded from a very wide range of habitats but analysis of the habitat data suggests a strong association with waste ground. However, there is no evidence it is an urban or suburban species despite the strong synanthropic tendencies quoted by Jeekel (2001) and by Kime (2001). It is often referred to as a pest in arable fields but less than 7% of the British and Irish records of the species were from cultivated land. The habitat analysis suggests a strong association with clay soils. Jeekel (2001) reported that the species seemed to prefer humid, humus-rich soils whether they were sands or clays. Blower (1985) and Kime (1995) both considered *B. superus* to reach the highest population densities on limestone, especially in woodland. Although below the threshold for a strong association, the habitat analysis does provide some evidence of a preference for calcareous soils. *Brachydesmus superus* is common throughout almost the whole of Europe and has been introduced to the USA, many of the Atlantic islands and Juan Fernandez Island in the Pacific (Kime, 2001). Adults have been recorded in all months of the year but most frequently from March to June. There are very few records of adults from August and September. It appears to be an annual species (Blower, 1979) that matures in the spring and dies following breeding / ovipositing. A few animals mature early and over winter as adults (Stephenson, 1961).

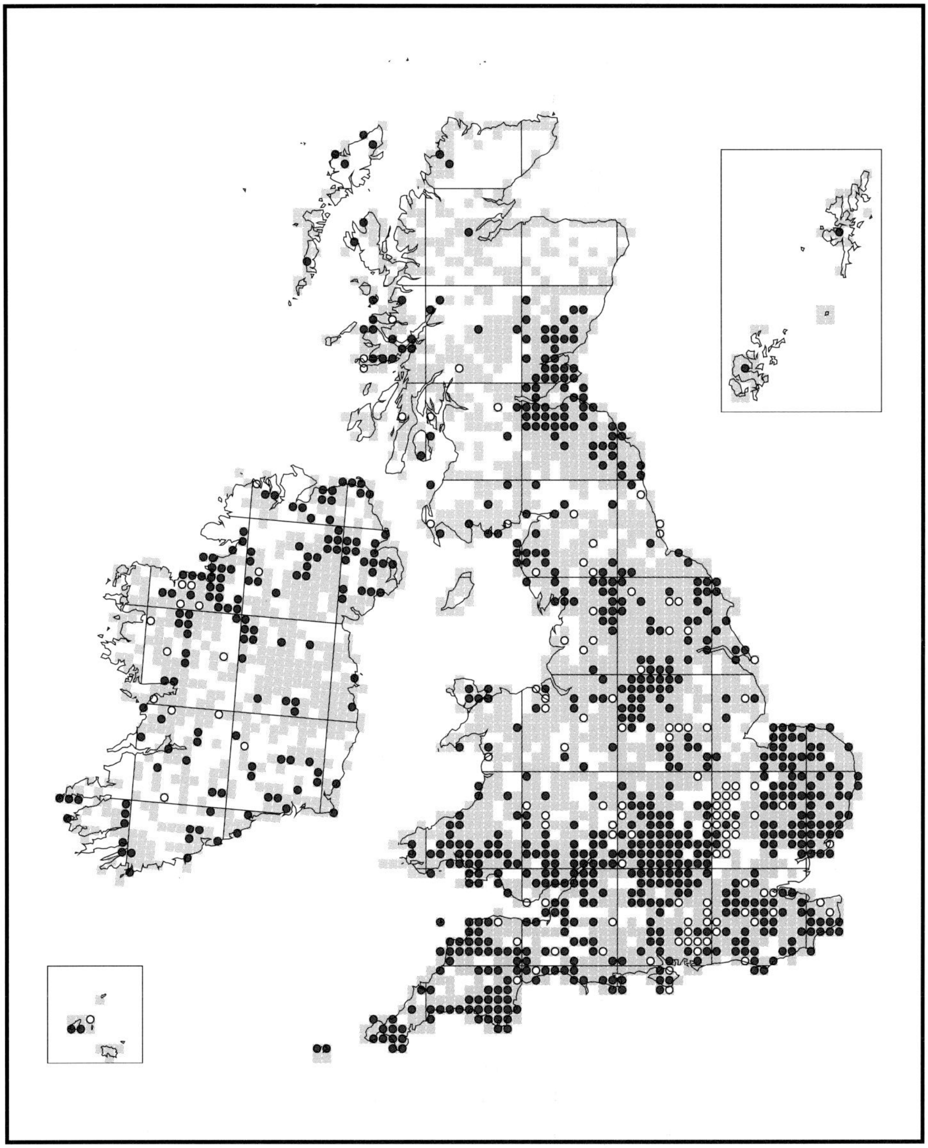

Polydesmus angustus

This is a common and widespread species in Britain and Ireland. It is very catholic in its requirements and is found in almost any habitat where millipedes occur. Many continental authors have reported it as a typical woodland animal (Haacker, 1968; Jeekel, 2001; Kime, 2001; Schubart, 1934) and it occurs frequently in British and Irish woodlands. However, analysis of the habitat data suggests a strong association with heath / moorland. It is the most frequently recorded species in this habitat which may explain an apparent association with peat soils. It may also account for a strong association with non-calcareous soils, something noted by Gregory and Campbell (1996) in Oxfordshire and by Kime (2004) in Belgium. However, Jeekel (2001) reported the species as occurring commonly on limestone in the Netherlands. He also found it frequently on clay but Kime and Wauthy (1984) linked it to soils with low clay content. In Britain and Ireland it has been recorded most frequently (45% of records) from loams. This contradictory information on soil preference probably reflects the eurytopic nature of the species. *Polydesmus angustus* has an extended Atlantic distribution, occurring as far east as Austria and the Czech Republic (Kime, 2004). It has been introduced to North America and Mexico (Blower, 1985). Work on a closely related species suggests animals probably take 30 to 36 months to reach maturity (Verhoeff, 1928). Both sexes die after reproduction, hence adults are scarce over the summer. However, there are some records of adults from all months of the year.

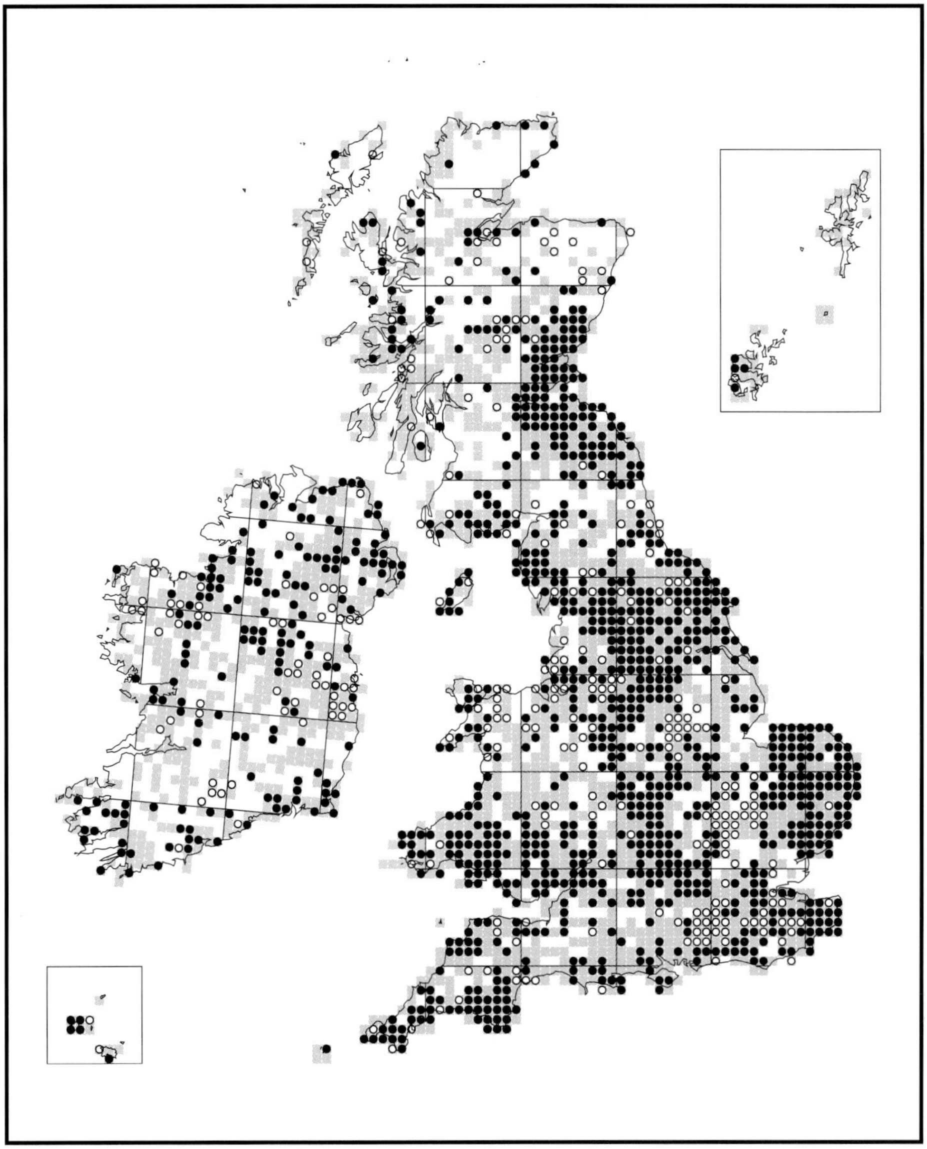

Polydesmus barberii

The first British record of this species was from Central Park, Plymouth in 1995. It has subsequently been recorded from other sites in and around Plymouth and in the Dartmouth area of south Devon. Although it was originally discovered in urban locations, *P. barberii* has also been collected from leaf litter in deciduous woodland. However, there are insufficient habitat data available for reliable conclusions to be drawn from any further analysis. In continental Europe it is found along the French and Italian Rivieras (Kime, 2001) and was presumably introduced to south Devon via the port or marina facilities in Dartmouth and Plymouth (Bolton, 1996). The species appears to be winter active with adults having been found in December, February and April. Immature stadia have been collected in November and December.

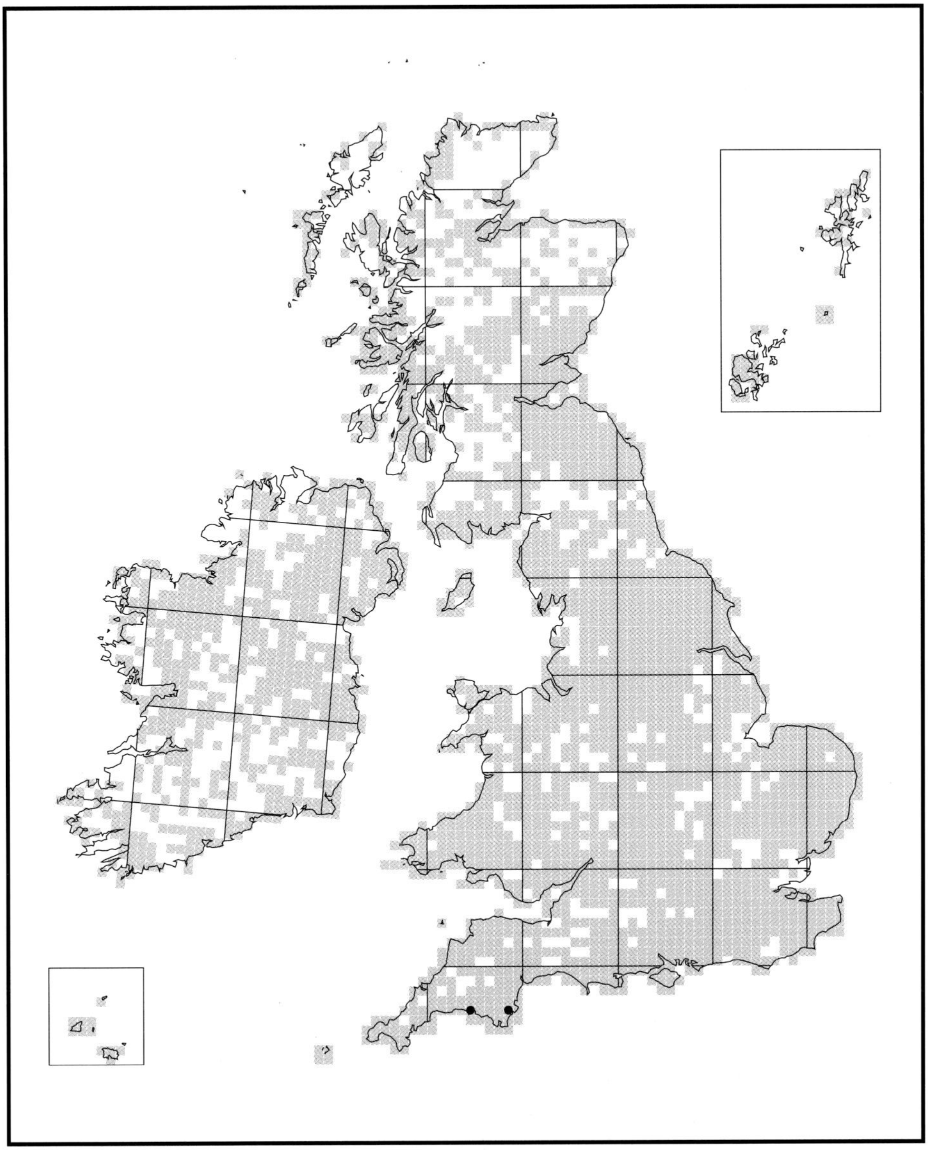

Polydesmus coriaceus
(= *Polydesmus gallicus*)

This is a common species in England, south of a line from the Humber to the Mersey, in Wales and in Ireland. It is uncommon in northern England and there is only one record from Scotland. It has been recorded in almost any habitat where millipedes occur but analysis of the habitat data suggests strong associations with cultivated land, waste ground and grassland. Kime (2004) reported it was common in a variety of habitats in northern Spain but associated it with wet conditions, even in more open habitats. Blower (1985) also noticed a link with damp situations. The habitat analysis shows a very strong association with clay soils and strong negative relationships with non-calcareous and sandy soils. Kime (1978) noted this preference for clay soils but not an aversion to non-calcareous soils. Van Den Haute (1999) found the species on sandy and loam soils in Belgium but Kime (2004) stated that it occurred on soils other than clay where restricted drainage or a high water table produced damp conditions. *Polydesmus coriaceus* has an Atlantic distribution extending southwards through Belgium and western France to Iberia, the Azores and Maderia. Britain and Ireland are at the northern limit of this distribution. This species has been introduced into the USA and the Cape Verde Islands (Kime, 2001). Kime (1995) noted that animals may be active all year round in milder parts of Britain and Ireland. Adults have been recorded in all months of the year but most frequently in the spring.

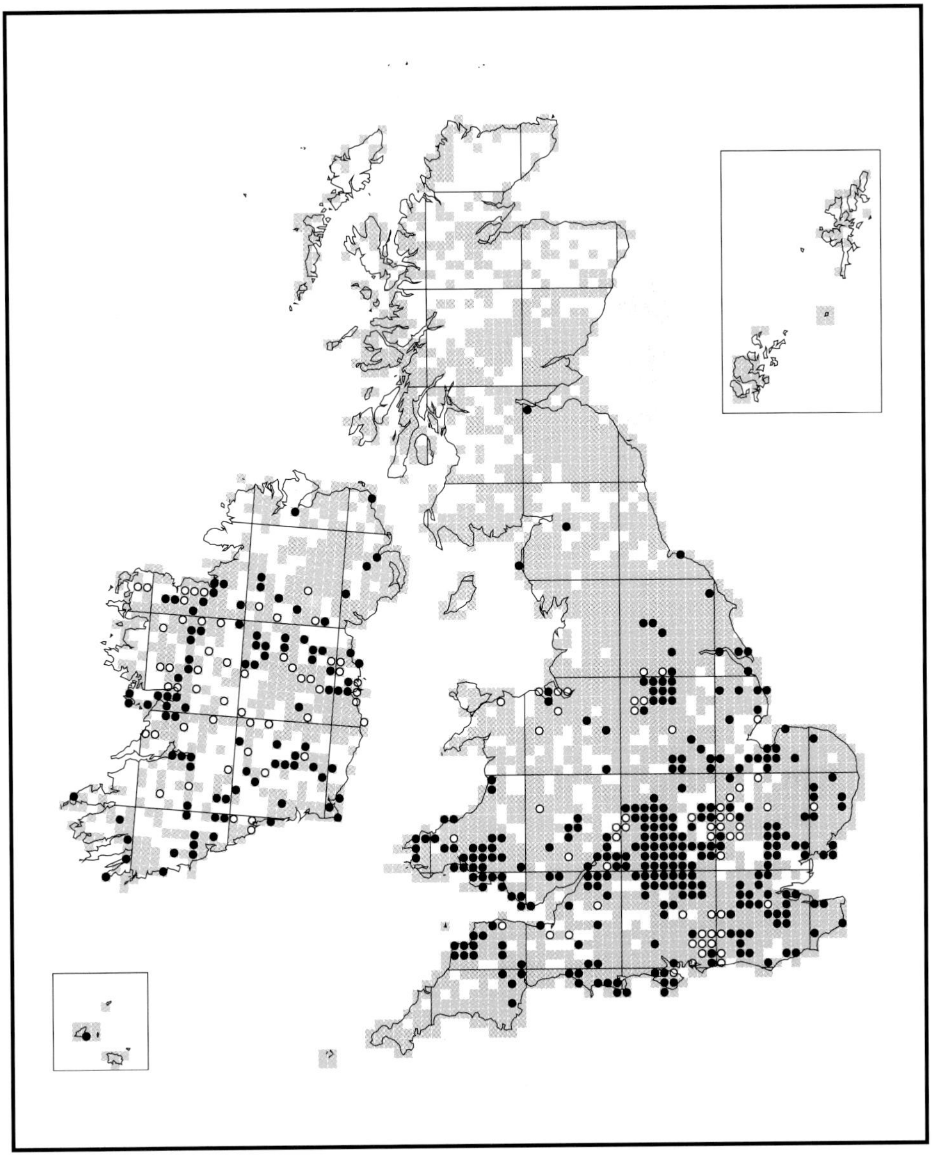

Polydesmus denticulatus

This is a frequent and widespread species in Britain and Ireland although records from Ireland and northern Scotland are relatively sparse. It has been found in a wide range of habitats, most frequently in woodland (55% of records) but analysis of the habitat data suggests a strong association with wetland. Blower (1985) considered it a woodland species and Kime (1995) reported it was usually found in wet woodland. In Belgium the millipede has also been found in a range of habitats but particularly in humid or wet conditions, often in woodland (Kime, 2004; Van Den Haute, 1999) and can survive inundation (Zulka, 1996). Further east in Luxembourg, Germany and Switzerland it is more typical of open habitats such as grassland and arable land (Kime, 1995; Pedroli-Christen, 1993; Schubart, 1934) whilst in the north and east of its European range it becomes synanthropic (Kime, 2001). In Britain and Ireland it shows a strong negative relationship with synanthropic urban and suburban sites. *Polydesmus denticulatus* is a widespread European species occurring from France eastwards to Russia and from Bulgaria and Italy north to Scandinavia. It has been introduced to Newfoundland (Blower, 1985). Adults have been recorded throughout the year but most frequently from April to September. Blower (1970) reported that animals from a population in a Cheshire woodland matured in their second summer and the peak activity of adults occurred in June and July.

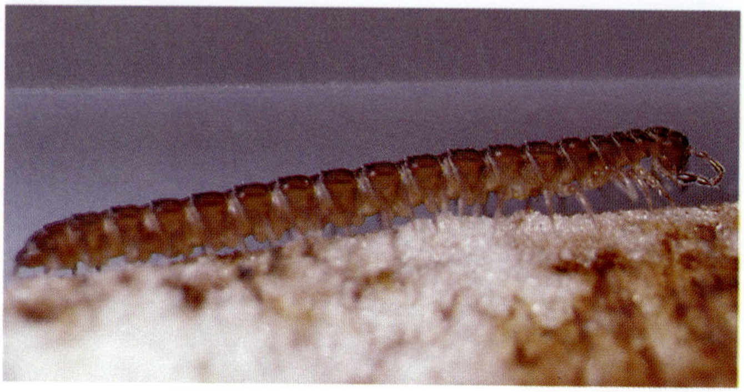

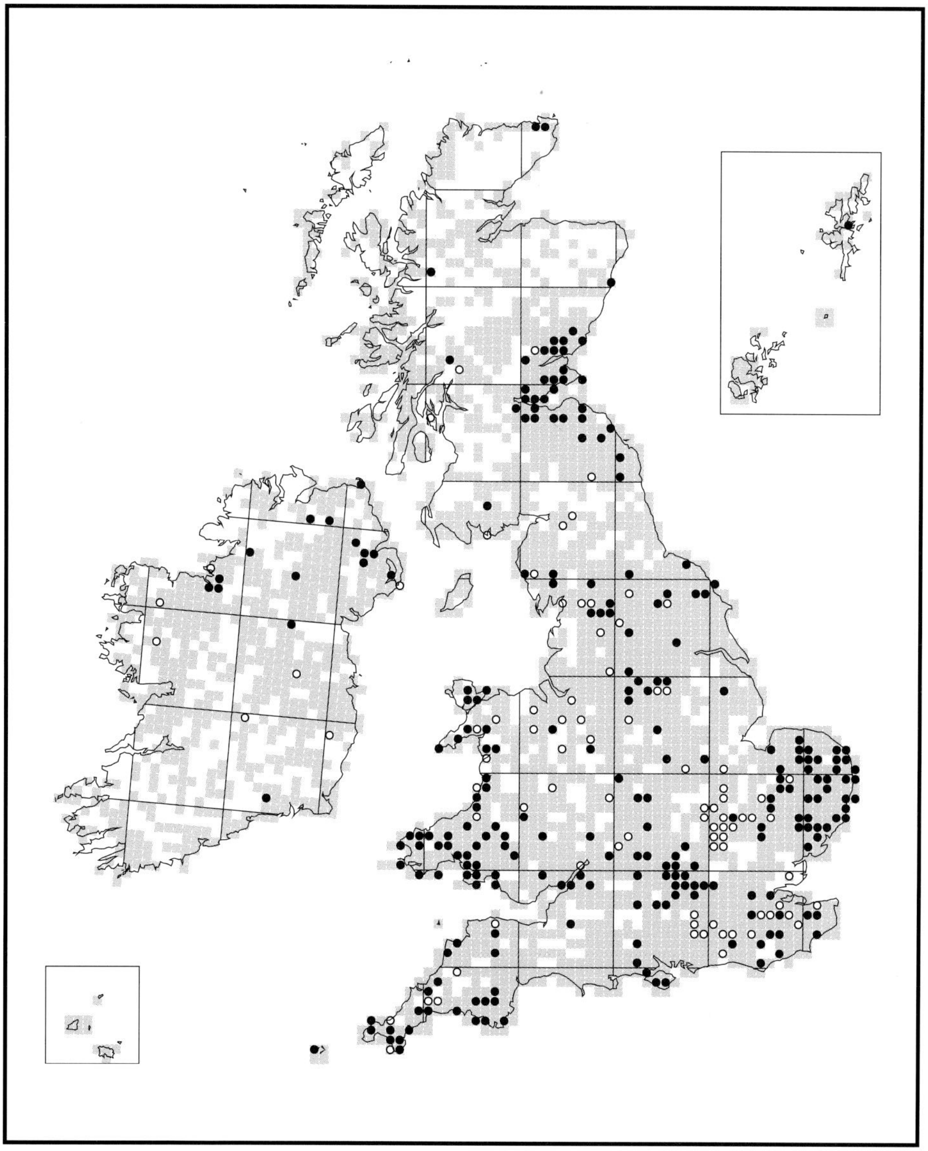

Polydesmus inconstans

This is a frequent and widespread species in Britain and Ireland. However, records from south west England, south west Ireland and Wales are surprisingly sparse considering the species has an Atlantic distribution. As with *P. angustus*, this species is found in almost any habitat where millipedes occur but analysis of the habitat data suggests strong associations with sand dunes and with grassland and a strong negative relationship with woodland. Kime (2004) reported it did not occur in sand dunes but was common in grassland in Belgium. Gregory and Campbell (1996) associated the species with damp river-side grassland in Oxfordshire. Although Kime (2004) noted it was especially common on calcareous grassland in Belgium, the habitat analysis provides no evidence of a preference for calcareous soils in Britain and Ireland. *Polydesmus inconstans* is found in western Europe from Spain to Norway, but has spread north as far as Iceland and east to the Ukraine through an ability to colonise synanthropic habitats (Kime, 1999). Jeekel (2001) suggested it may even be a recent arrival in the Netherlands. Blower (1985) stated it had been introduced to North America. Observations on a population of this species in Michigan, USA show animals from eggs laid in summer will mature the following spring (Snider, 1981) but Blower (1985) suspected temperatures in Britain and Ireland are probably too low to allow this annual life cycle. Both sexes die following reproduction so few adults are present from late summer through to the following spring when a new generation matures. Although there are a few records of adults from all months of the year, they are collected most frequently between April and June.

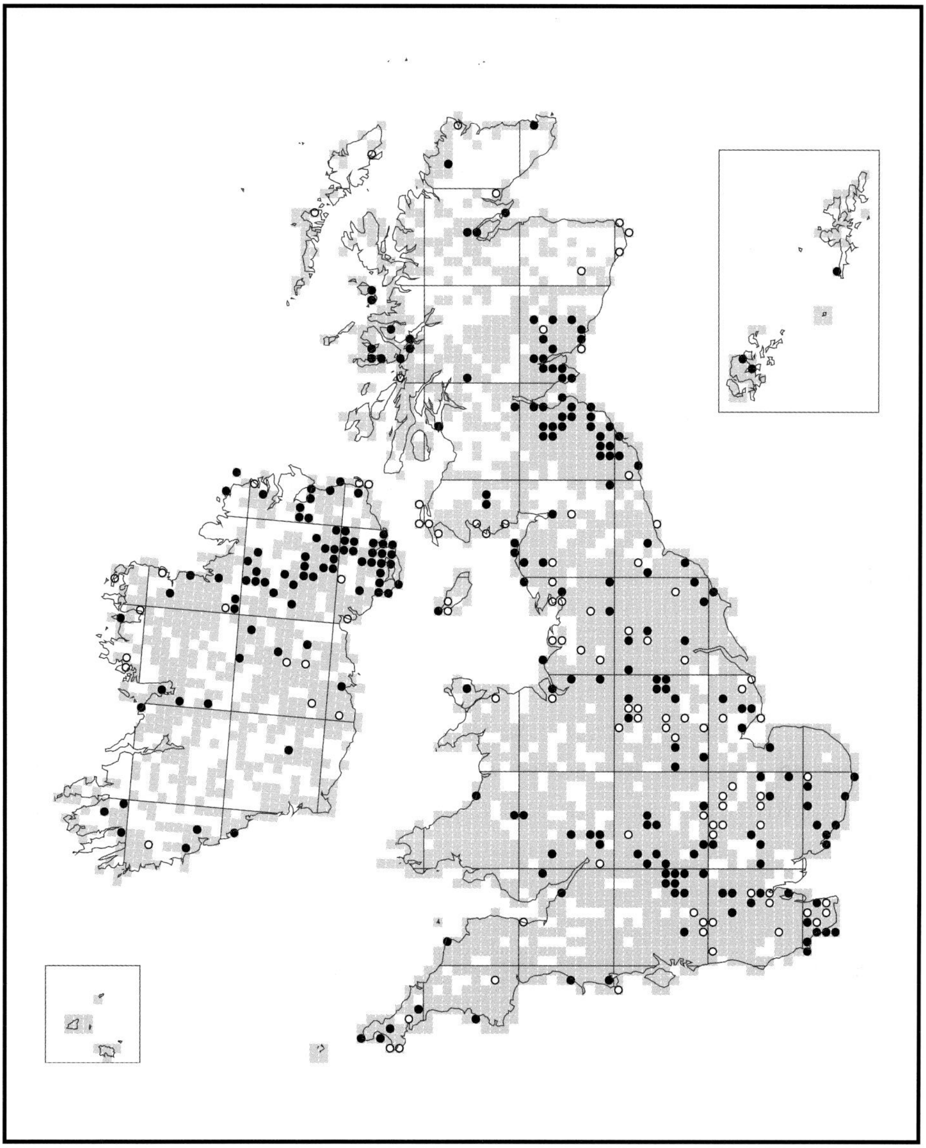

Propolydesmus testaceus
(= *Polydesmus testaceus*)

The first British record of this species was from Walton-on-the-Naze in Essex (Pocock, 1903). It has not been found in Essex since, nor in Cornwall where it was later noted (Turk, 1944). Most modern records are from Kent with other records from chalk downland in Oxfordshire (Gregory & Campbell, 1996), reclaimed industrial land and woodland in Monmouthshire (Harper, 2004) and most recently from woodland in Suffolk (Lee, 2005c). It has been recorded from a range of habitats but there are insufficient ecological data available for reliable conclusions to be drawn from any analysis. Most European records are from open habitats but the species has also been found in caves and damp woodland and Kime (2004) considered it to be thermophilic with a preference for calcareous soils. This is in general agreement with the British data; most of the records are from relatively open habitats but there are records from caves and woodland and all are from base-rich soils. *Propolydesmus testaceus* has an extended Atlantic distribution from south west France to the Czech Republic and the species appears to reach the northern limits of its range in Britain and Germany (Kime, 2001). In contrast with its scarcity in Britain, Jeekel (2001) found the species to be common in the limestone area of Limburg in the Netherlands. Kime (2001) also reported collecting large numbers of specimens by pitfall trapping in Belgium and Luxembourg. In Britain adults have been collected in each month from April to July and again in October.

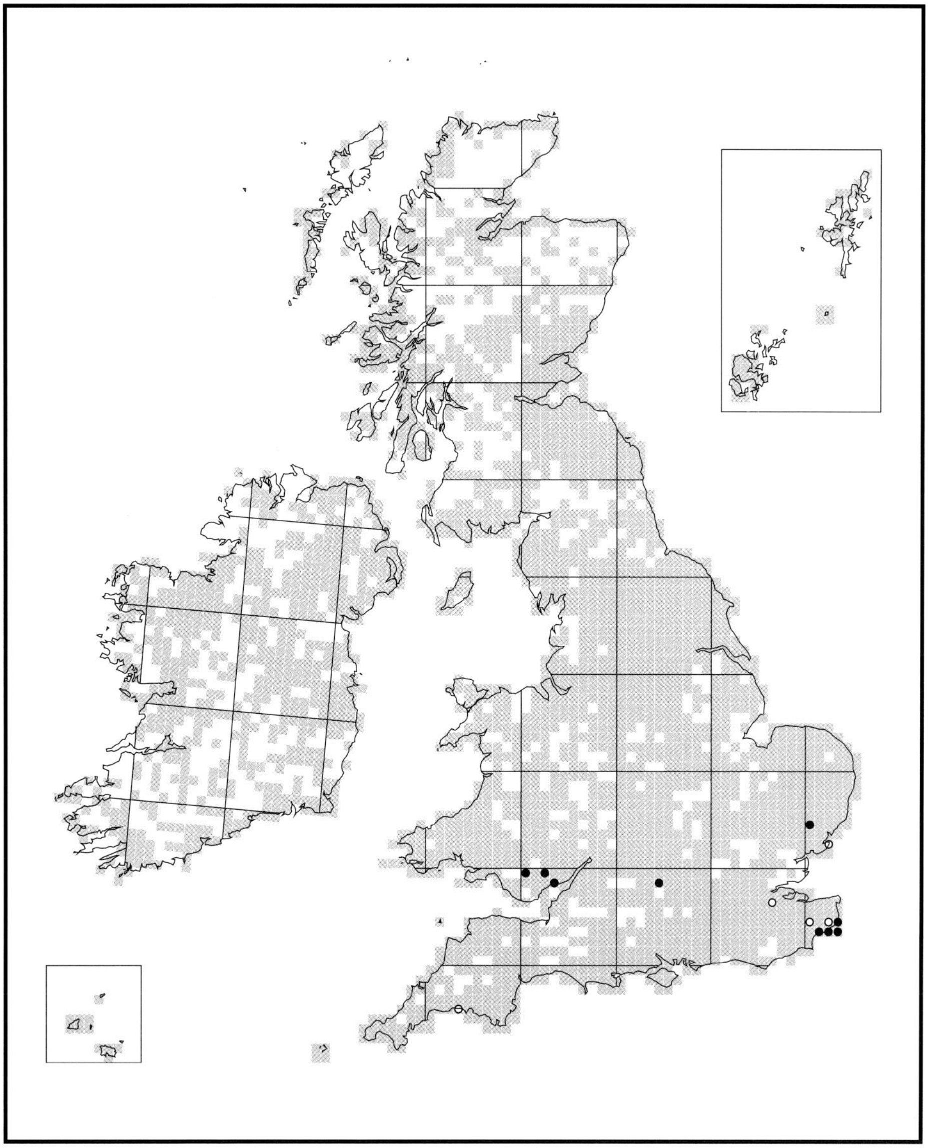

Macrosternodesmus palicola

This species has proved to be widespread in Britain and Ireland. It is apparently absent from much of northern Scotland, north Wales and south west England. This, in part at least, may be related to the underlying geology as *M. palicola* is usually described as a calcicole. Kime (1995) reported its occurrence in beech woods on the chalk downs of southern England whilst Gregory and Campbell (1996) reported its presence in ancient deciduous woodland and synanthropic sites on calcareous soils in Oxfordshire. Analysis of the habitat data suggests it is very strongly associated with synanthropic suburban / urban sites especially churchyards and cultivated land such as allotments, gardens and parks. In these synanthropic habitats it is not restricted to calcareous soils and has been collected from gardens on acid sands in East Anglia and gritstone in Yorkshire. In Belgium it is mainly found in woodland on chalk and limestone with only a few records from synanthropic sites (Kime, 2004). *Macrosternodesmus palicola* has an Atlantic distribution extending from the Pyrenees to Norway. It occurs as a synanthrope further east in Germany and Sweden (Kime, 2001). In Britain and Ireland it has been recorded frequently in the winter months from the underside of large stones, often during periods of sharp frosts. Adults have been collected in each month from October to July, most frequently in April and May.

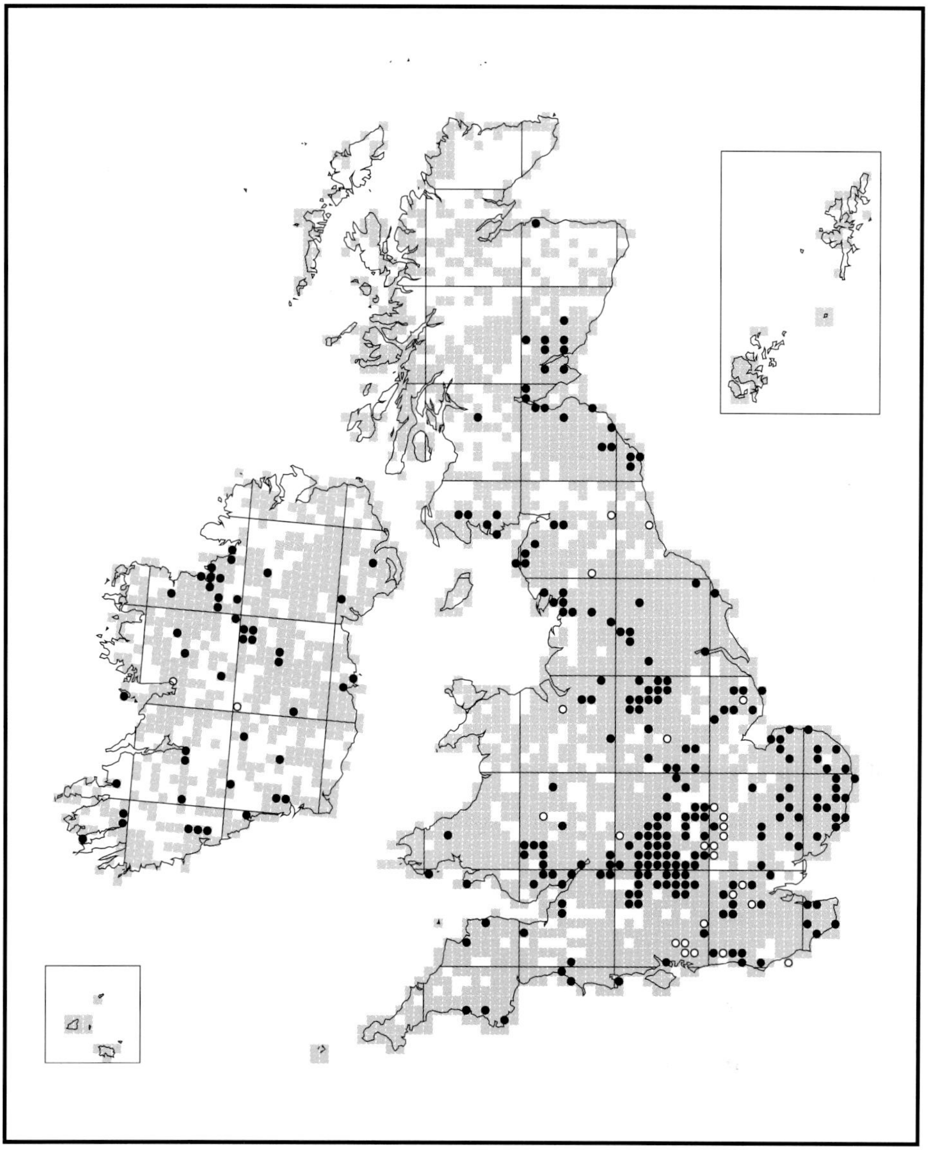

Ophiodesmus albonanus

Although recorded a little less frequently, this species has a very similar distribution in Britain and Ireland to that of *Macrosternodesmus palicola,* with which it is often found. Thus it is absent from much of northern Scotland, north Wales and south west England. Analysis of the habitat data suggests that, like *M. palicola,* this species is very strongly associated with synanthropic suburban / urban sites especially churchyards and cultivated land such as allotments, gardens and parks. Most records (78%) are from sites on calcareous soils but, like *M. palicola,* it also occurs on more acidic soils in synanthropic habitats. The situation is similar in continental Europe where it has an Atlantic distribution but occurs at a much lower frequency than *M. palicola.* Jeekel (2001) considers the species to be a rarity in the Netherlands. It has been found as far east as Poland and as far north as Sweden in synanthropic habitats (Kime, 2001). It has also been introduced to Newfoundland (Blower, 1985). Gregory and Campbell (1996) report it is easiest to find in the spring in Oxfordshire. In Britain and Ireland adults have been reported from February to November but most frequently in April and May.

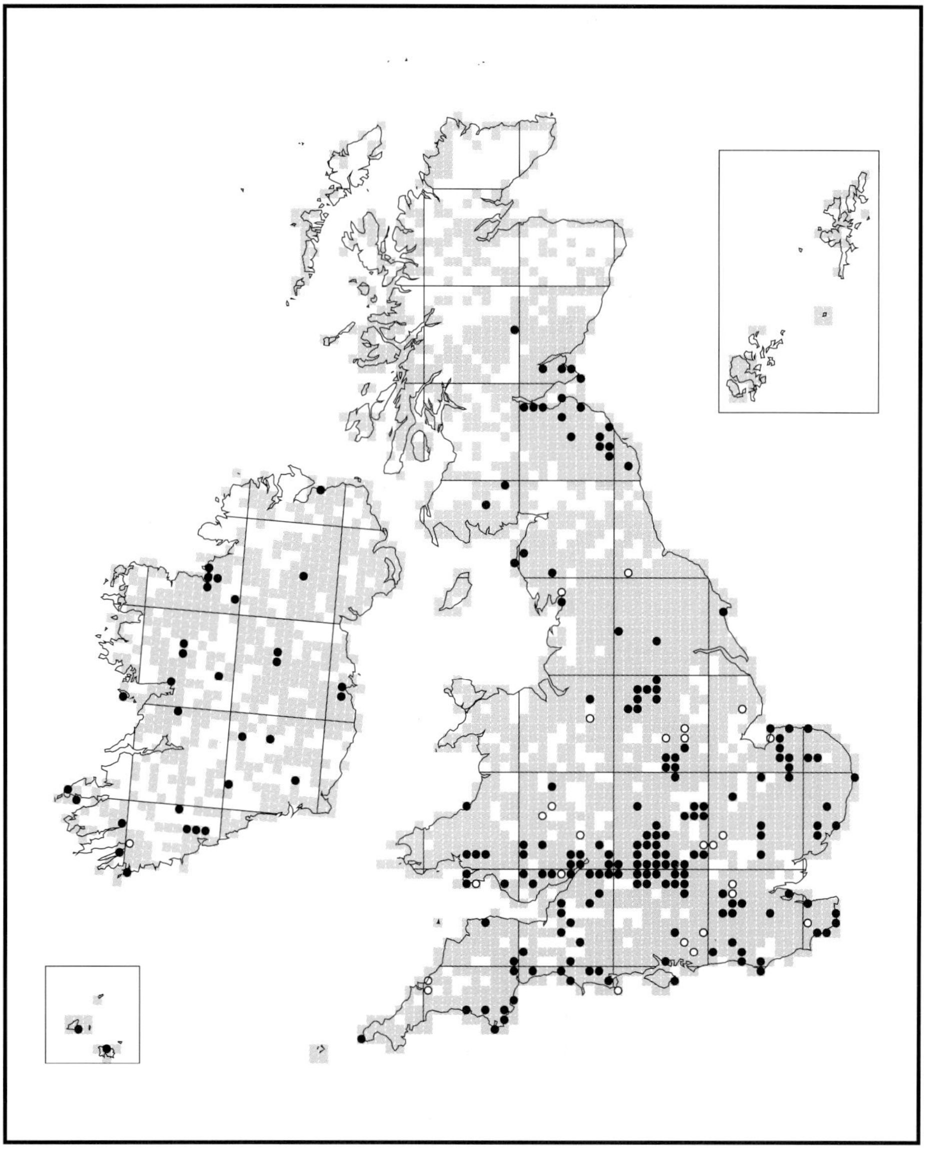

Choneiulus palmatus

This is a widespread but uncommon species in Britain and Ireland. In Scotland it has not been recorded from north of Montrose. Analysis of the habitat data suggests a very strong association with buildings and with urban sites. This synanthropic tendency is strongest in the north of England and in Scotland where the species is usually recorded from gardens and glasshouses. Further south it can be found in the same habitats but has also been collected from leaf litter or under bark in deciduous woodland. Kime (1999) suggested this might be its natural habitat. Pedroli-Christen (1993) also reported the species from calcareous grassland in Switzerland. Kime (2004) noted that many of the semi-natural sites where this species occurs are on calcareous soils. There are insufficient data available from Britain and Ireland to check the validity of this statement. The species appears to be native to the Atlantic zone of north west Europe but occurs as a synanthrope further north and east and has been introduced to Madeira, the Azores and North America (Kime, 1999). Adults have been found from February to June and from September to November. They probably occur throughout the year. The life history of this species has not been studied in detail but it is likely to be similar to that of *Proteroiulus fuscus*.

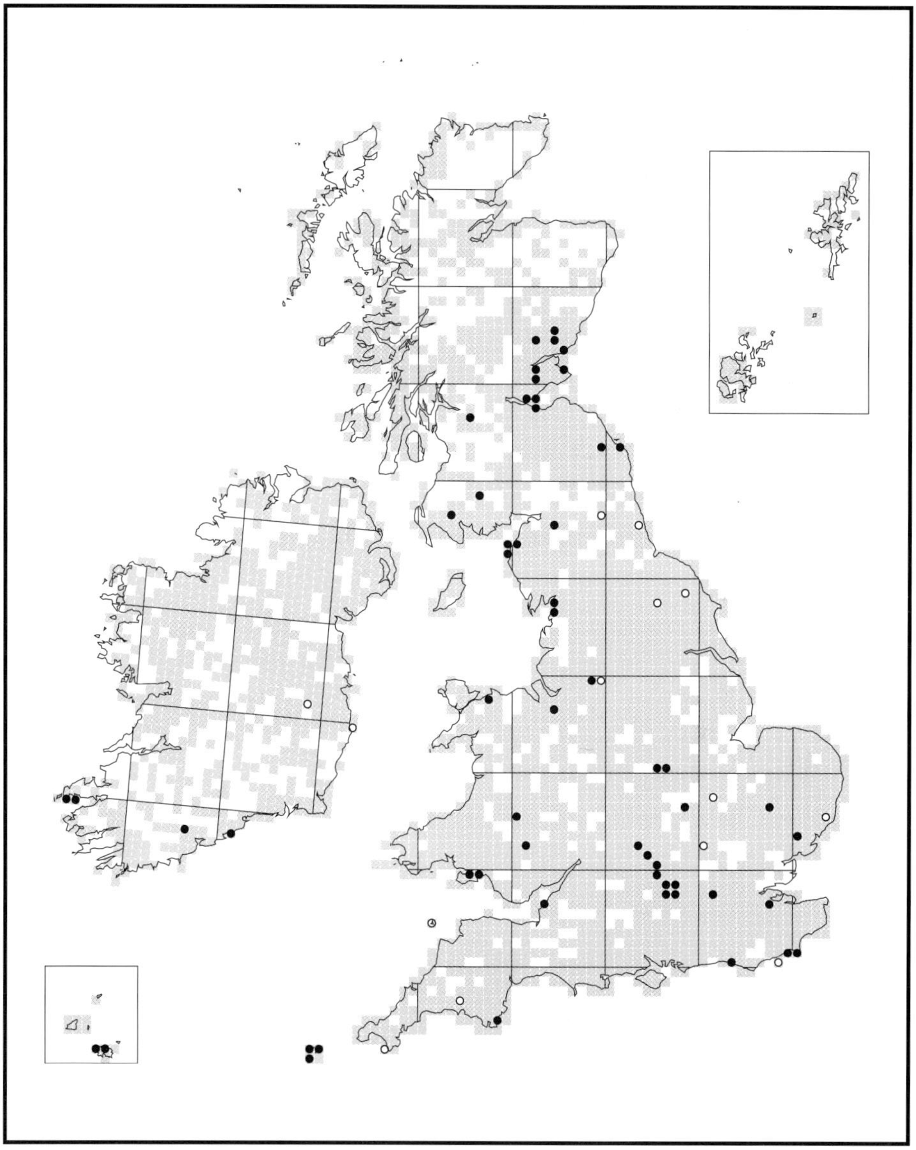

Nopoiulus kochii

British specimens have been referred to this species since at least 1911 but there has been a great deal of confusion surrounding the nomenclature of *Nopoiulus kochii* and related species. Where it has been possible for modern workers to re-examine these early specimens they have all proved to belong to other taxa, namely *Blaniulus guttulatus*, *Choneiulus palmatus* and *Proteroiulus fuscus* (Blower, 1985). The first confirmed specimen of *N. kochii* was not collected in Britain until 1986 (Hopkin & Blower, 1987) but it has since been collected from a small number of locations scattered across Britain and Ireland. There is relatively little habitat data for this species but the analysis suggests that it is highly synanthropic showing strong associations with buildings and with waste ground. The millipede is strongly synanthropic in continental Europe but it has been collected from caves (Kime, 2004) and from grassland (Pedroli-Christen, 1993). It may have originally been a central European species but now is recorded widely across the continent and has been introduced to New Zealand, North and South America (Kime, 1999). Adults have been found from April to June and from October to January. They probably occur throughout the year. The life history of this species has not been studied in detail but it is likely to be similar to that of *Proteroiulus fuscus.*

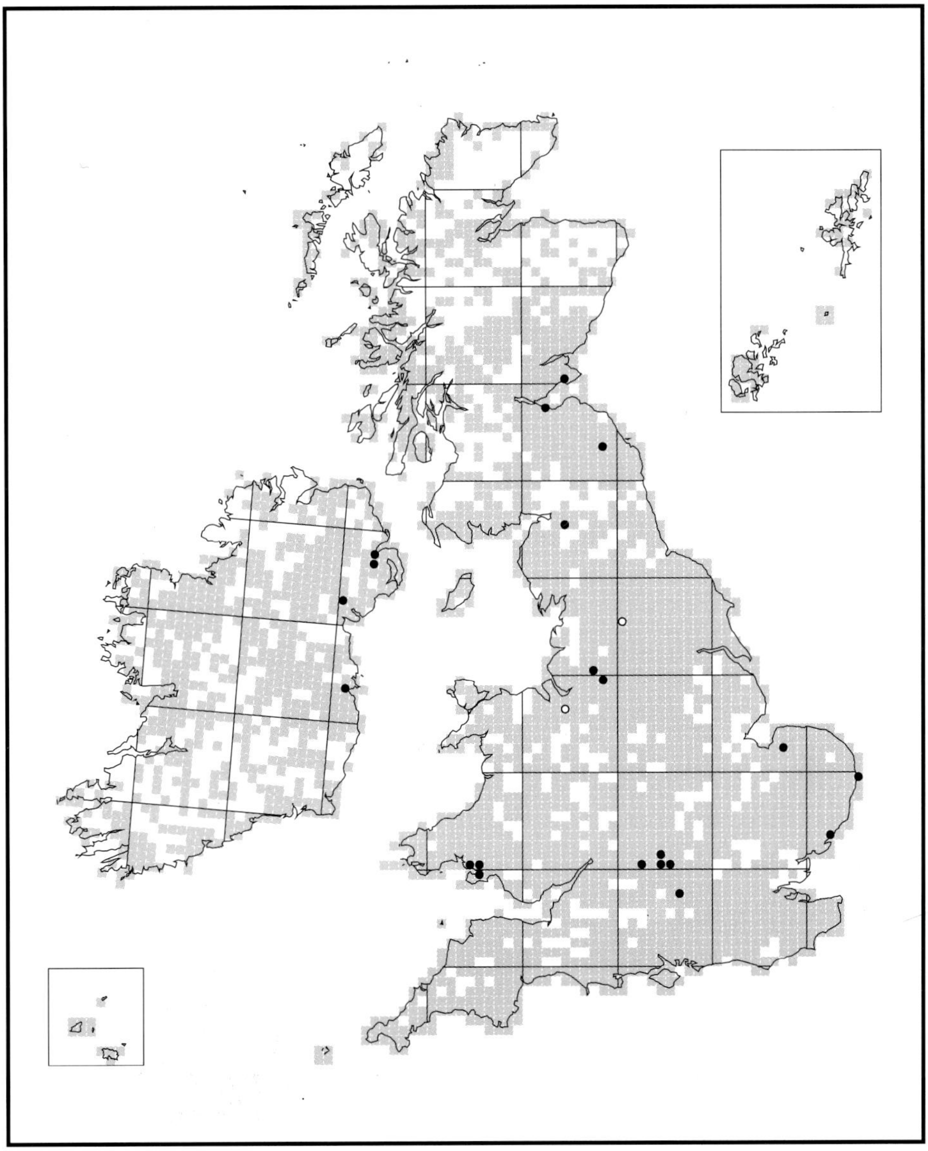

Proteroiulus fuscus

This species is very widespread and common in Britain and Ireland. Analysis of the habitat data suggests a very strong association with woodland and strong negative relationships with synanthropic sites including churchyards, gardens and waste ground. Strong negative relationships with arable land, grassland and sand dunes are also evident. These findings reflect the fact that, although not restricted to woodland, *P. fuscus* is usually found under bark on both living and dead coniferous and deciduous trees. It can also occur in soil and leaf litter, either in well established populations (Blower, 1985) or during dispersal (Kime, 2004). Analysis of the habitat data suggests a strong association with non-calcareous soils in general and peaty soils in particular. It is essentially a northern European species and is restricted to mountains in the south of its range. It appears to be absent from much of France, Iberia, the Balkans and the Mediterranean region. However, it has been introduced to North America, South Africa and a number of islands in the Atlantic (Kime, 1999). As with *Nemosoma varicorne,* this is a parthenogenetic species and it is rare to find males. Where males do occur they appear to be non-functional (Enghoff, 1978). Females take three years to reach maturity in natural conditions (Brookes, 1974) and adults may be found at any time of the year. In laboratory conditions a male has survived for a further six years on reaching maturity (Rantala, 1974).

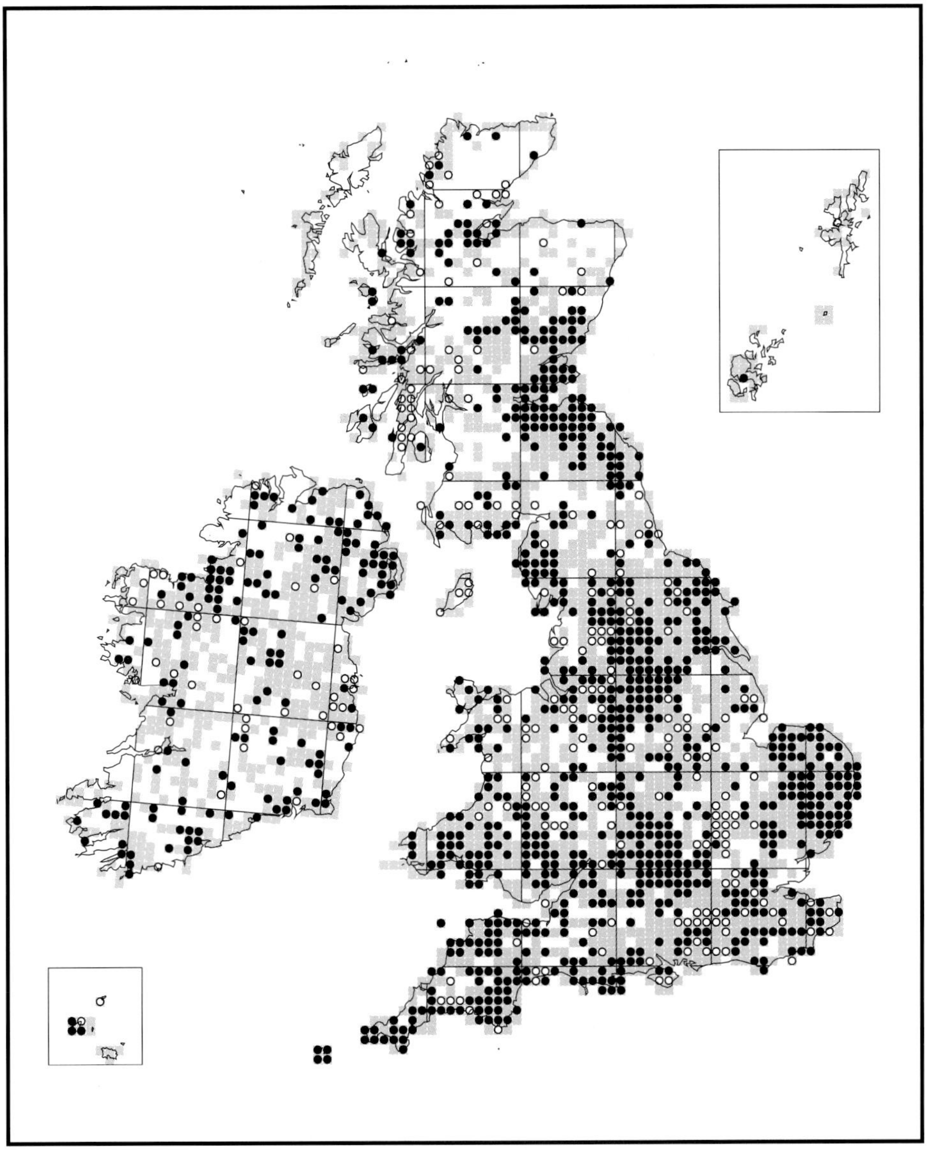

Blaniulus guttulatus

This is a widespread and common species throughout much of Britain and Ireland but is apparently scarce in the extreme north west of Scotland. Analysis of the habitat data suggests a very strong association with cultivated land and with synanthropic sites including gardens and churchyards. A strong association with waste ground is also suggested. Despite being the most frequent habitat category recorded for this species, the analysis suggests a strong negative relationship with woodland. In contrast Kime (2004) considered it may be a natural component of the woodland fauna on calcareous, loamy soils in Belgium. Although not strong, analysis of the environmental data does provide some support for a link with calcareous loams. In Europe the species is widespread in the Atlantic zone and north central Europe but there are only scattered records further south and east (Kime, 1999). It is a synanthropic species that has been introduced to North America and islands in the Atlantic Ocean (Blower, 1985) and is well known as a pest in agricultural and horticultural systems. Brookes and Willoughby (1978) found that the young millipedes take four years to reach maturity. Adults have been found throughout the year but most frequently in the spring and again in autumn. They are harder to find in winter as falling autumn temperatures cause them to move deeper into the soil until temperatures rise again at the end of winter (Pierrard, Bonte & Baurant, 1963).

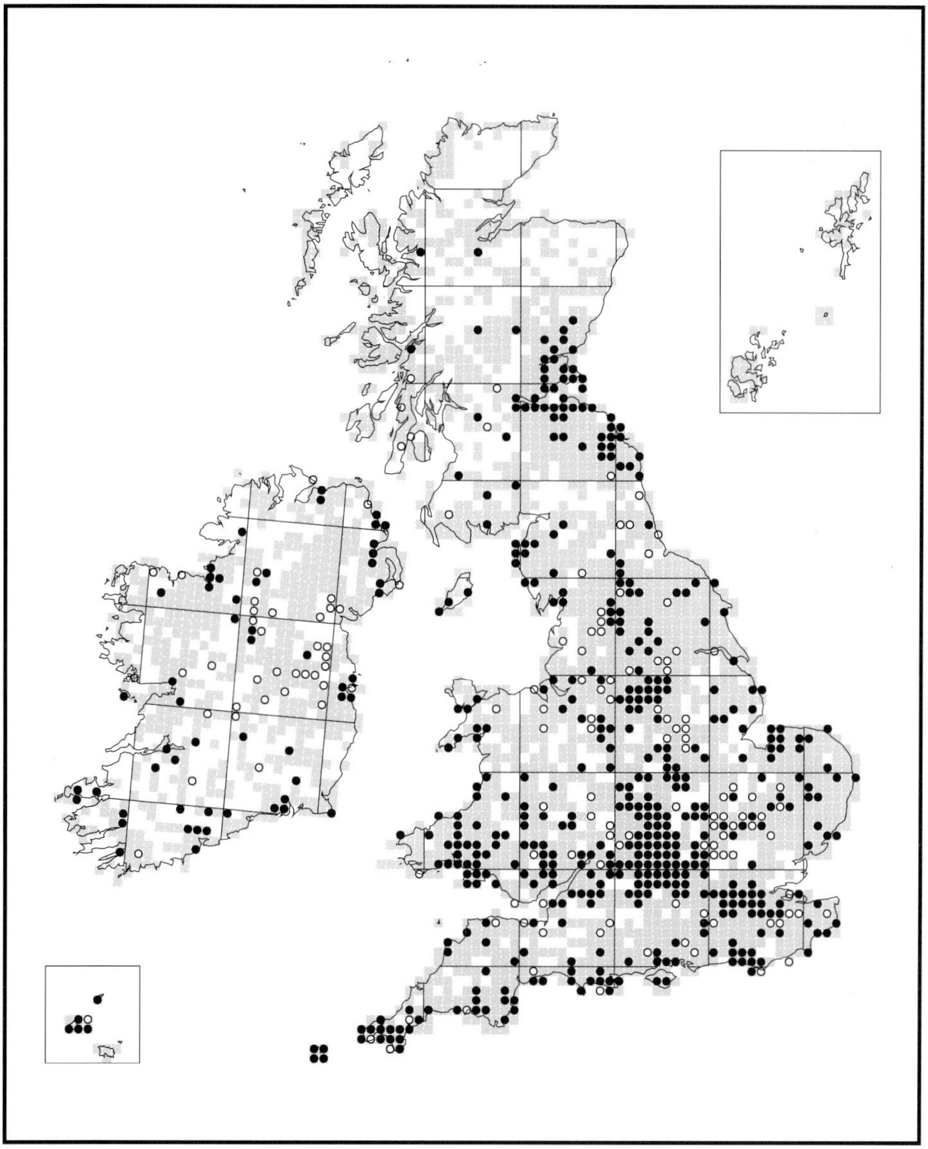

Archiboreoiulus pallidus

This species is widespread in Britain but appears to be scarce in the west and in Ireland. There are no records of it from the mainland north of Angus but it is recorded from Orkney. Analysis of the habitat data suggests a strong association with cultivated sites and with urban / sub-urban locations suggesting this is another garden species. However, it has been found in a variety of habitats including woodland, arable farm-land and grassland. Elsewhere in Europe the species has been found on calcareous soils in arable land, forests, grassland and synanthropic sites and also in caves (Kime, 1999, 2004). Blower (1985) associated the species with calcareous soils in Britain but analysis of the environmental data does not support the existence of such a link. Kime (1999) describes the continental distribution of this species as extended Atlantic with a scattering of records extending eastwards to Russia in synanthropic sites. The species has also been found in Canada (Blower, 1985). Adults have been found throughout the year but most frequently in the spring and again in autumn. They are harder to find in winter as falling autumn temperatures cause them to move deeper into the soil until temperatures rise again in the spring (Biernaux & Baurant, 1964). Although hand searching and pitfall trapping will result in the collection of surface active animals, the use of soil extraction techniques are likely to be necessary to gain a true picture of the distribution and abundance of this species.

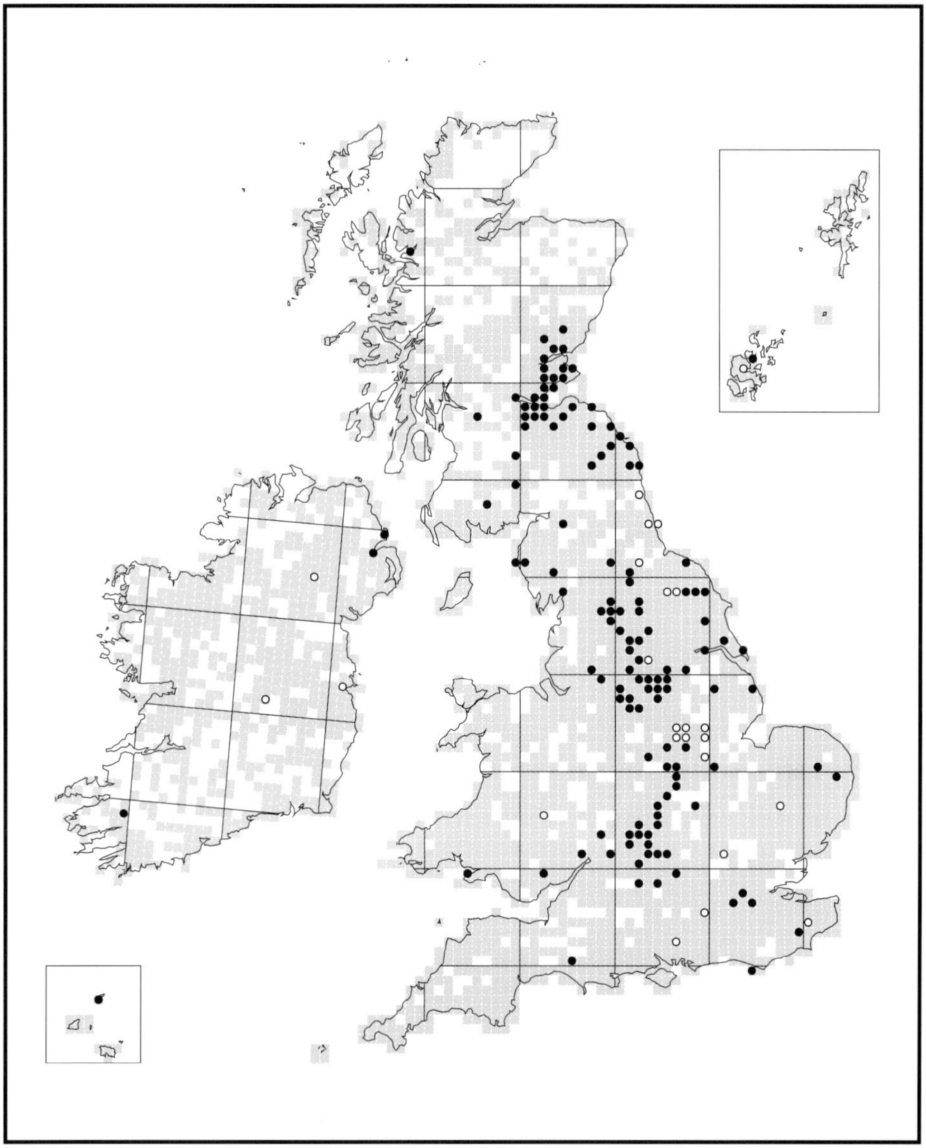

Boreoiulus tenuis

This species is widespread in Britain and Ireland but there is a surprising paucity of records from southern England and Wales. Although the clusters of records in certain areas clearly originate from active field workers there are equally active collectors in areas where the species has never been recorded. Blower (1985) recognised *B. tenuis* as a species of arable fields and it may become a pest in sugar beet (Baker, 1974). Kime (2004) considered the species highly synanthropic and analysis of the habitat data supports this. A strong association with cultivated land and particularly with synanthropic sites including gardens and churchyards is apparent. Blower (1985) also stated that this species is associated particularly with woodland and with base-rich soils. However, although 30% of records of this species are from woodland, the habitat data provides no evidence of an association with either woodland or with calcareous soils. It is a central European species occurring as far eastwards as Lithuania but only just reaching the north and east of France. It has a distinct northern bias to its distribution as it occurs in southern Scandinavia but not in Austria, the Czech Republic, Slovakia or anywhere further south (Kime, 1999). Adults have been found throughout the year but most frequently in the spring. As with other blaniulids, surface activity is likely to be determined by temperature and soil moisture. Surprisingly, Gregory and Campbell (1996) noted that, in Oxfordshire, the species appears to be easier to find during winter.

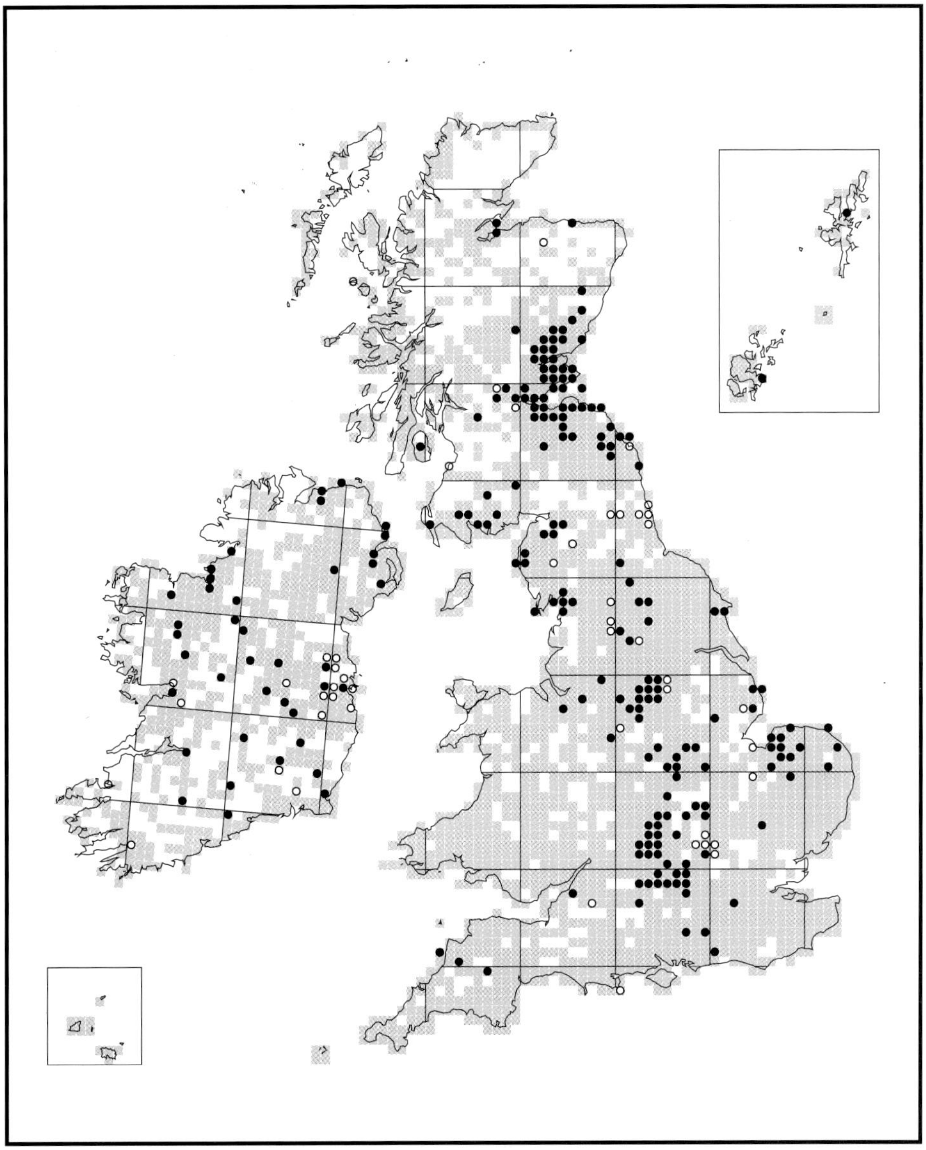

Nemasoma varicorne

Although widespread in Britain and Ireland, this species is less frequently collected than *Proteroiulus fuscus* with which it is often found. Enghoff (1976b) collected the species from beneath the bark of 20 different genera of coniferous and deciduous trees. As might be expected for a sub-cortical species, analysis of the habitat data suggests a strong association with woodland. Insufficient data were available to detect any preference for particular tree-species but the millipede shows a preference for beech in Denmark (Enghoff, 1976b) and Belgium (Kime, 2004), for aspen in Sweden (Lohmander, 1956) and for alder in Germany (Haacker, 1968). *Nemasoma varicorne* has an essentially Central European distribution (Kime, 2004). Enghoff (1994) has shown that normal bisexual populations are found only in the centre of its range and are replaced by a parthenogenetic form in the peripheral regions including Britain and Ireland in the west, southern Scandinavia in the north and Russia in the east. Thus male specimens are only rarely collected in Britain and Ireland and are probably non-functional. Adults have been found throughout the year but most frequently in April and May. Maturity is achieved after two years and many animals then reproduce again the following year but very few survive beyond this stage (Brookes, 1974). This species is usually collected by searching under bark on dead trees and stumps but can also be taken in bark traps. Kime (1997) has taken the species in pitfall traps in grassland during the winter and suggested this was the season when it dispersed.

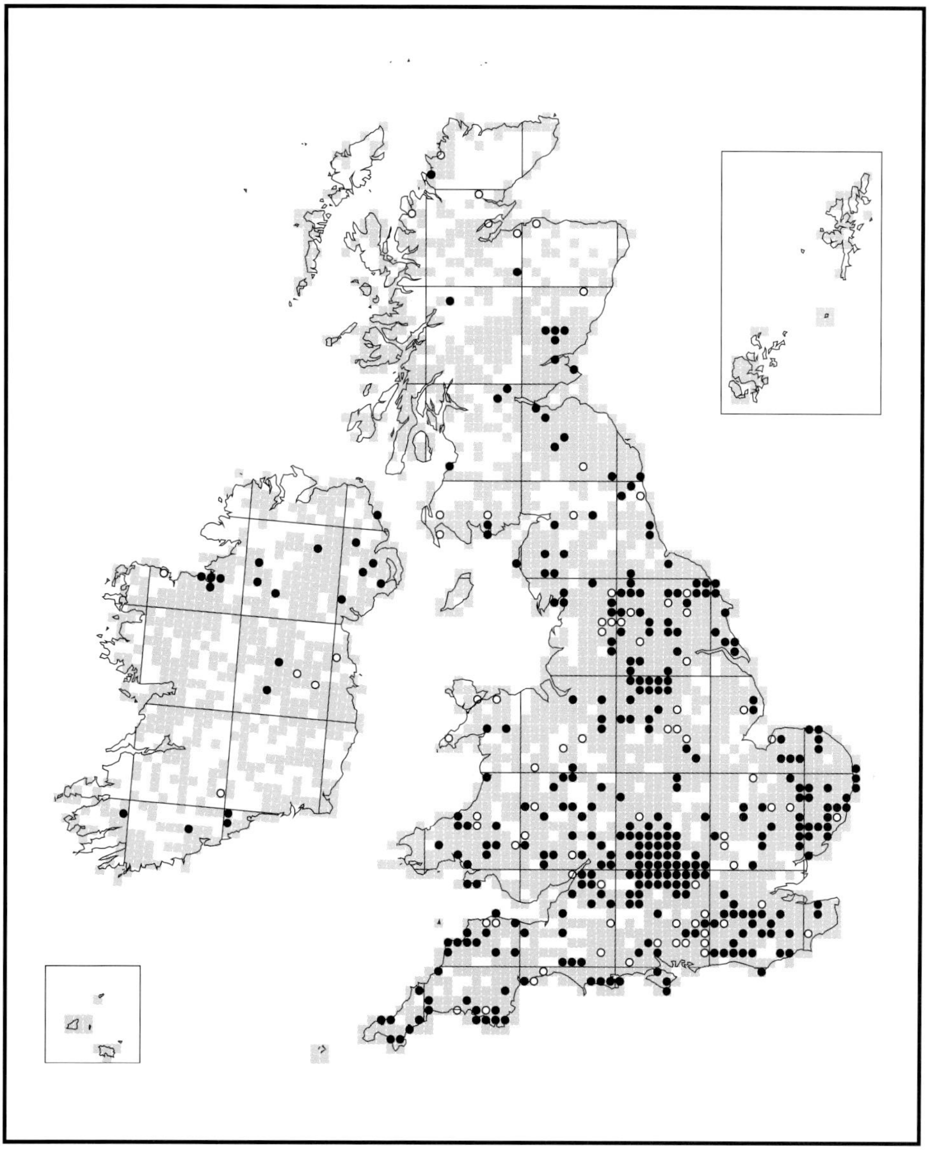

Thalassisobates littoralis

This species has been collected from between the tide marks at a scattering of coastal locations around Britain and Ireland. Harding (1985) mentioned a record from Blakeney Point on the Norfolk coast but this is not shown on the map as a precise grid reference was not available. Analysis of the habitat data suggests a strong association with maritime habitats as would be expected for the only truly littoral millipede found in Britain and Ireland. It appears to be associated with shingle or rocky coasts (Kime, 1999) but has not been reported from the major shingle formations of East Anglia (excepting Blakeney Point) and the south coast of England (excepting Slapton Ley). With the exception of a single record from Sweden, all other European records of this species are from the western Mediterranean (Kime, 1990a). Enghoff (1987) reports its introduction to North America leading Kime (1999) to question whether it has also been introduced to Britain and Ireland from the Mediterranean. Details of the life history of this species have still to be determined but adults have been collected in April, June and August. It can be difficult to collect this millipede by hand even at sites where it is known to occur. Some workers have had success using pitfall traps in shingle (Harding, 1985) but this technique has failed to produce specimens in apparently suitable locations at Orford Ness, Suffolk and Dungeness, Kent.

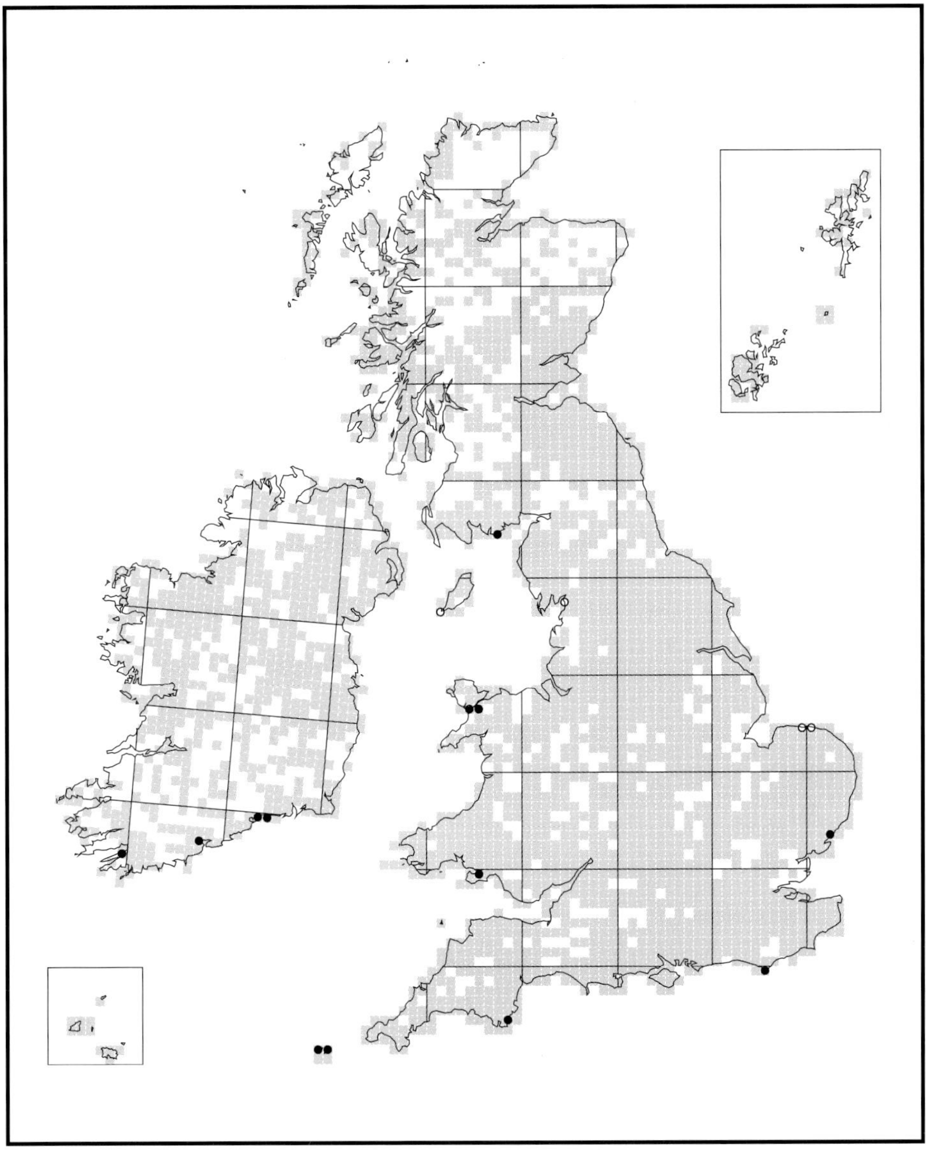

Julus scandinavius

This is a widespread species across Britain and Ireland but, although common in much of the country, it is surprisingly scarce in the extreme south east and central southern England. It has been collected from a wide range of habitats including grasslands, wetlands and sand dunes but analysis of the habitat data suggests a strong association with woodland. It is chiefly a woodland species in continental Europe (Kime, 2004; Pedroli-Christen, 1993). It appears to be less synanthropic than *Ophyiulus pilosus* and shows a strong negative relationship with cultivated land and waste ground. Blower (1985) states that it has a preference for more acid soils and, although it does occur in calcareous grassland, Kime (1995) reports it is never found in limestone woodlands in Europe. Although below the threshold for a strong association, the habitat analysis does suggest a preference for non-calcareous soils. *Julus scandinavius* is also a common species in central Europe but it is absent from most of France including the channel coasts (Kime, 1999). It has been introduced to the USA (Jeekel, 1973). Animals take three years to reach maturity but after breeding in the spring both sexes die and hence there are few adults around by late summer (Blower, 1979). However, there are some records of adults from all months of the year.

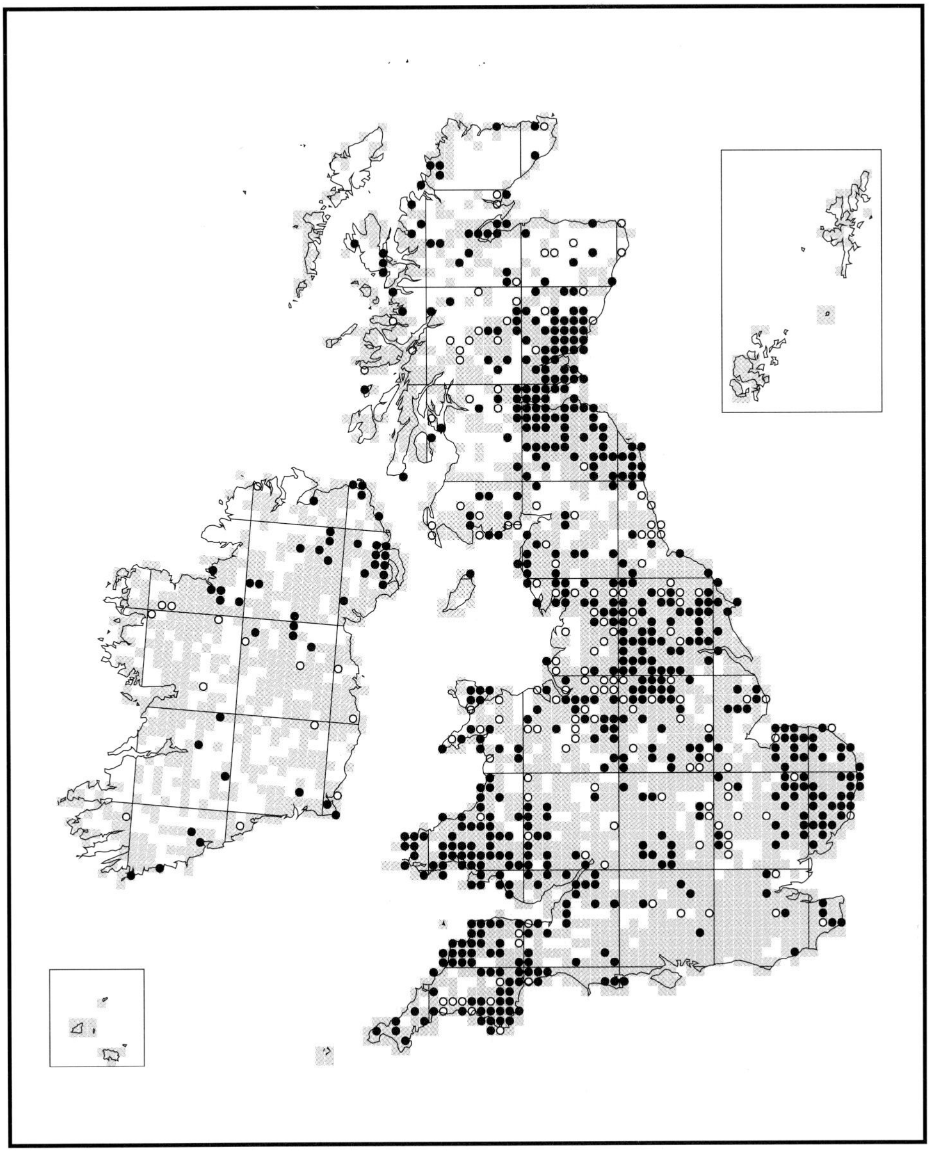

Haplopodoiulus spathifer

Although originally collected from Kew Gardens in 1976, this species was not identified from Britain until 1986 (Corbet & Jones, 1996). Subsequently it has been recorded from just two further sites, Bedgebury Arboretum in Kent and Wakehurst Place in East Sussex. It is found outdoors rather than in glasshouses but there are insufficient records available for reliable conclusions to be drawn from further analysis of the recording scheme data. Corbet and Jones (1996) reported that it occurred under stones and dead wood but preferred deep, deciduous leaf litter that retains moisture through the summer. This species has a strict Atlantic distribution and is native to a small region of south west France and north west Spain centred on the western Pyrenees (Kime, 1999). The millipede must have been introduced to Kew Gardens and probably spread from there to Bedgebury and Wakehurst Place, both of which have connections with Kew. The millipede can clearly overwinter in southern England and it is surprising that it does not appear to have spread to areas local to the botanic gardens. Adult animals have been collected in March and June.

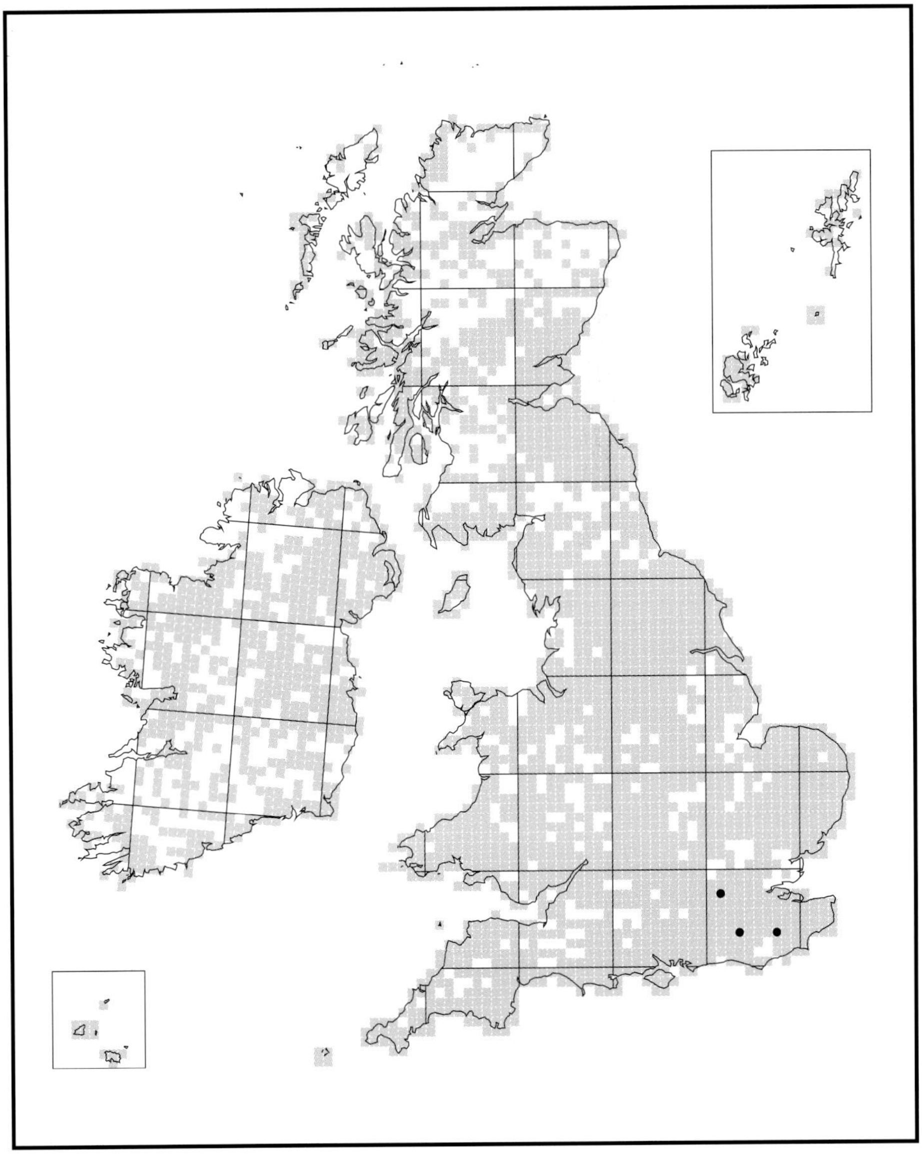

Ophyiulus pilosus

This is one of the commonest and most widespread species in Britain and Ireland. It was considered typical of leaf litter in deciduous woodland by Blower (1985) and 45% of the records of this species were from woodland. However, it has been found in a very wide range of habitats and analysis of the data suggests it has no strong association with any particular habitat. The analysis does provide some evidence of the synanthropic tendencies of this species, especially on cultivated land. Kime (1999) suggested that *O. pilosus* may have been introduced to Britain and Ireland from central Europe rather than having spread naturally following the last glaciation. This view is based on the fact that it is common in north Italy and nearby alpine areas but is rare or absent in much of northern Europe, with the exception of southern Scandinavia where it could also have been introduced. Jeekel (2001) agreed that *O. pilosus* was probably introduced to Britain but considered it native to Scandinavia and thought that this was probably the source of the British and Irish populations. The species has also been introduced to New Zealand and North America (Blower, 1985). Animals take two years to reach maturity but after breeding in the spring both sexes die and hence there are few adults around by late summer (Blower & Miller, 1974). However, there are some records of adults from all months of the year.

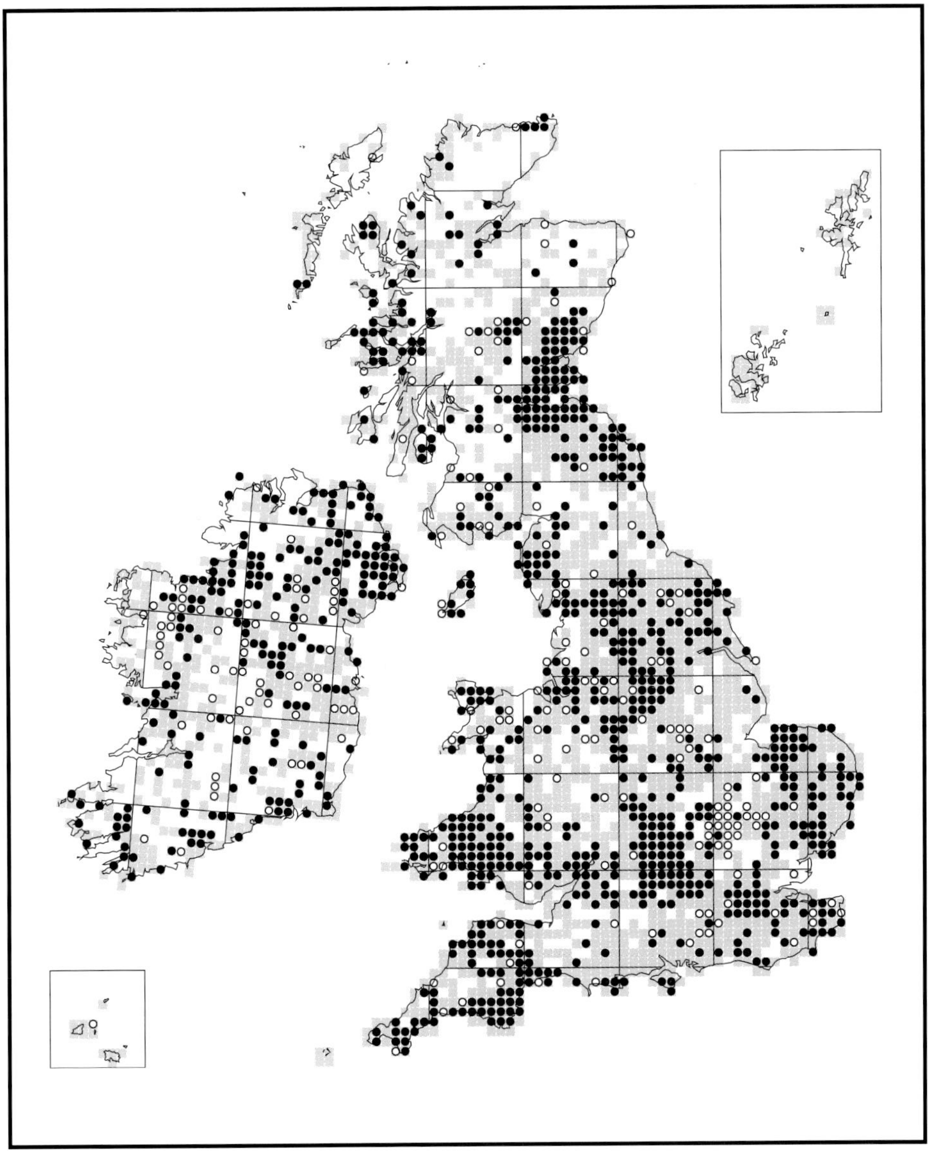

Leptoiulus belgicus

In Britain this species is most frequent in south west England and south Wales but there are isolated records from Eigg, Kent, Norfolk and north Wales. Surprisingly it has not been recorded from the south west of Ireland; the few Irish records are all from the northern coasts. It has been collected from a wide range of habitats and analysis of the data suggests it has no particular preferences although Morgan (1988) linked it with synanthropic habitats in south west Wales. Kime (2004) listed a wide range of habitats for *L. belgicus* in Belgium but associated it with warm positions on light, well drained soils. The habitat analysis provides no evidence of a preference for particular soil types but it does suggest a strong association with coastal sites and this is probably due to the warmer microclimates in these areas. This species has an extended Atlantic distribution suggesting it ought to be more widespread in southern England. It is found throughout western Europe as far east as Germany and the western tip of Austria but reaches its northern limit in Belgium (Kime, 1999). Animals are active throughout the year but show a pronounced peak in activity in late summer and autumn when they mature (Kime, 1999; Morgan, 1988; Pedroli-Christen, 1993). It is likely that the life history is similar to that of *Julus scandinavius* (Blower, 1985). Thus breeding would occur in the spring. Both sexes then die so few adults should be expected during the summer months.

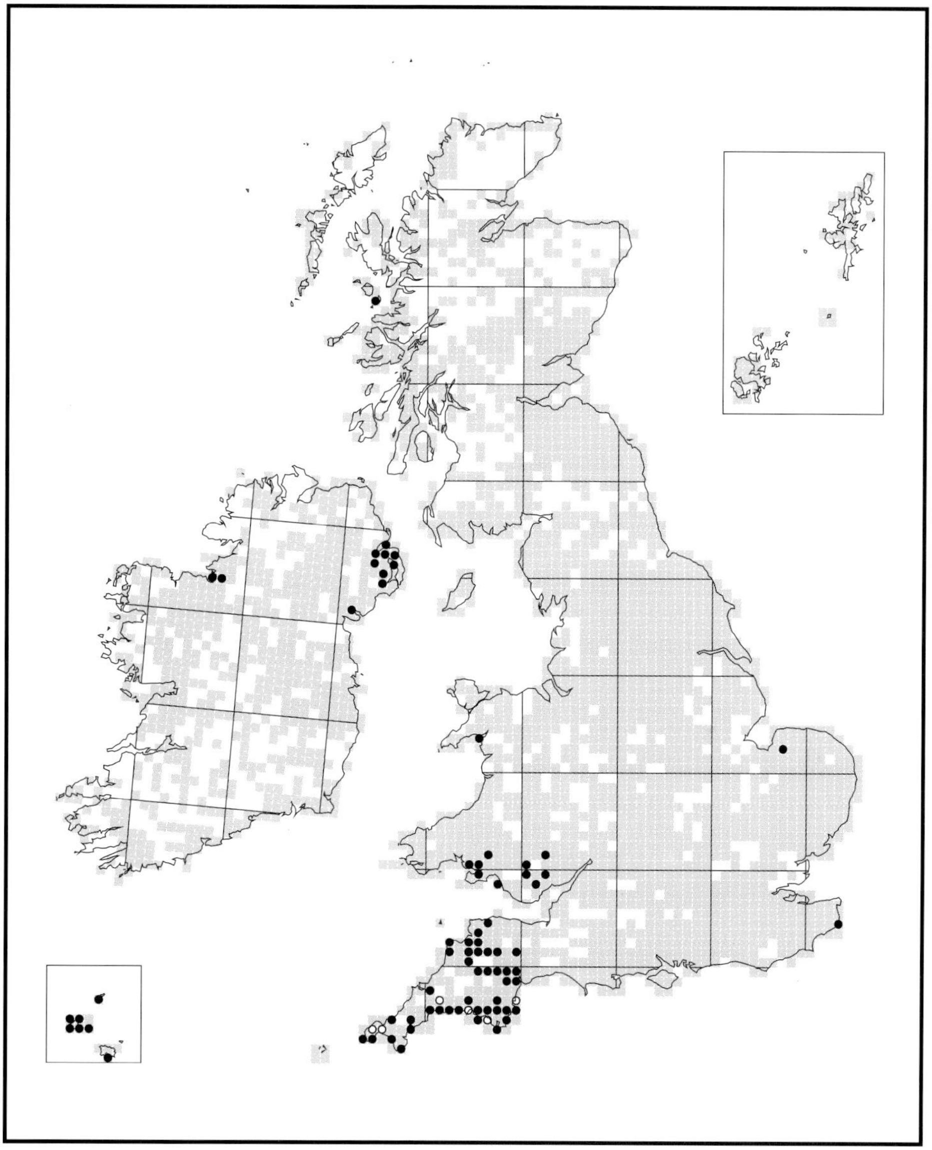

Leptoiulus kervillei

Although most frequent in Devon and Cornwall this species has been recorded from several counties south of a line from the Severn to the Thames estuary and from south Wales. Blower (1985) refers to a record from Bedford but no details of this were submitted to the recording scheme. Kime (2004) describes it as a common woodland species in the milder parts of Belgium. He associates it with base-rich soils, high in silt content. Analysis of the habitat data shows no strong relationships with any of the habitats or environmental factors considered. However, there is some evidence it shows a preference for loamy, non-calcareous soils in coastal habitats. As with *L. belgicus*, coastal areas probably provide the warm conditions favoured by this millipede at the northern limits of its range in southern England. Kime and Wauthy (1984) found that this species preferred areas in Belgium with the highest average temperatures and the mildest winters. There is no obvious explanation for the apparent contradiction with the Belgian data on the millipede's preference for soil acidity levels. *Leptoiulus kervillei* has an Atlantic distribution, occurring through northern and western France and reaching its northern limit in the Low Countries (Kime, 1999). It is likely that the life history is similar to that of *Julus scandinavius* (Blower, 1985). Thus breeding would occur in the spring. Both sexes then die so few adults should be expected during the summer months. Adults have been recorded from January to July but the number of records involved is small.

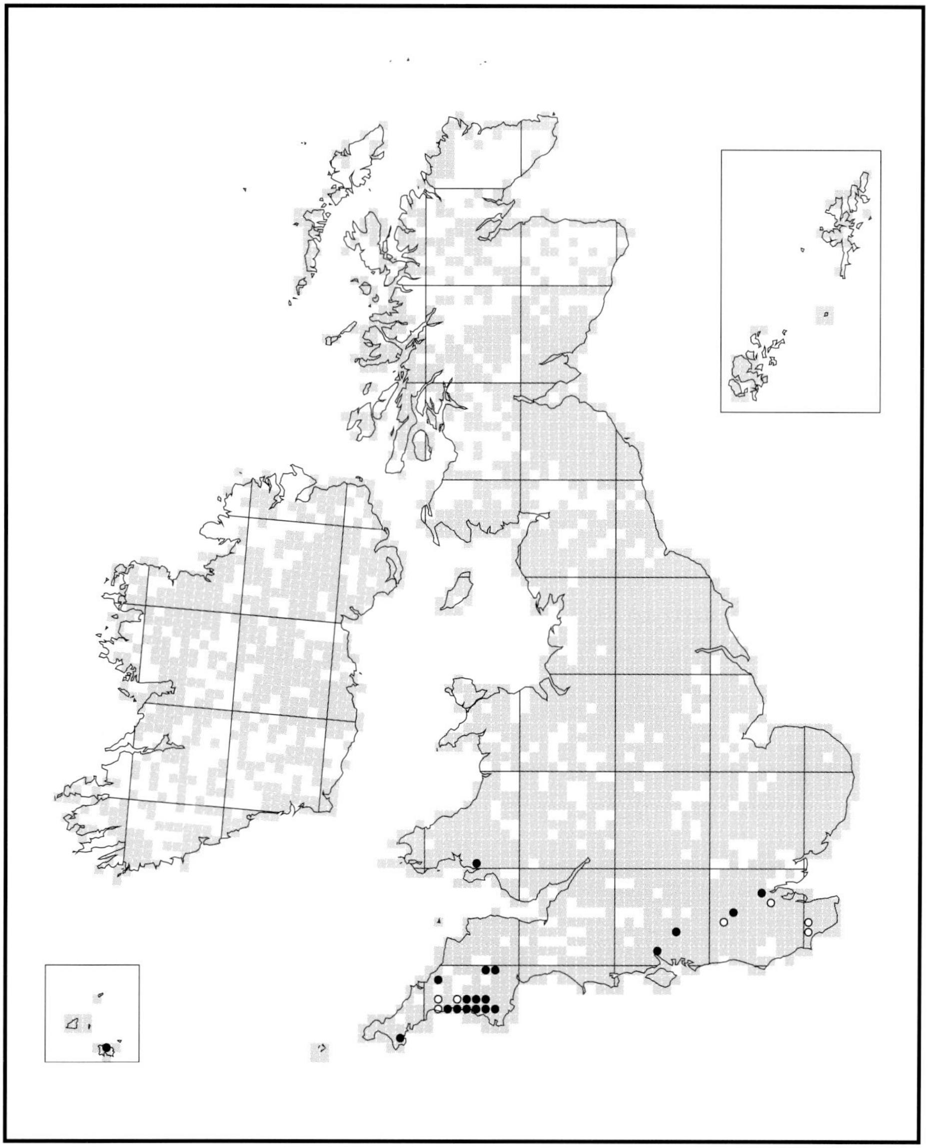

Metaiulus pratensis

This species was described as new to science from specimens collected in Kent and Sussex between 1939 and 1951 (Blower & Rolfe, 1956). It remains restricted to the extreme south east in England and appears to have undergone a severe decline even within this limited range over the last 25 years. Habitat details were submitted to the recording scheme with just two records hence there were insufficient ecological data for analysis. However, many of the records of this millipede are from arable land on heavy clay soils (Blower 1985). Most recently, between 1986 and 1988, it was found in a wetland and in four woodlands along the valley of the River Medway in Kent (Eric Philp, pers. comm.). It has been collected from farmland in France (Kime, 1999) but also from three caves by Demange (1958, 1965) and Mauriès (1965). *Metaiulus pratensis* has a strict Atlantic distribution occurring only in south east England and south west France on current knowledge. Thus the British populations of this millipede are of international importance. Little is known of its life history. Adults of both sexes have been collected in June and in each month from August to November. An adult male has also been collected in January.

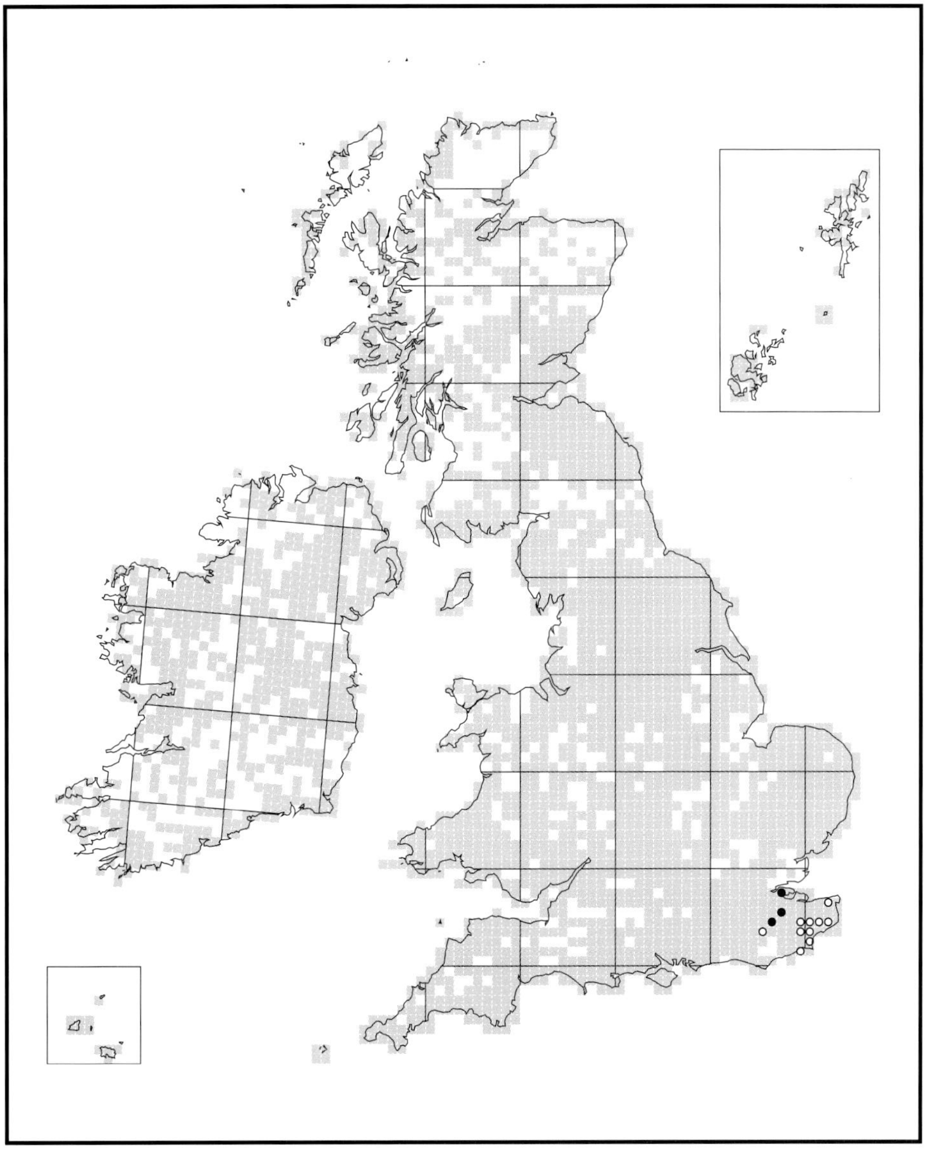

Allajulus nitidus
(= *Cylindroiulus nitidus*)

This is a widespread but uncommon species that appears to be more frequent in the east of Britain. It has not been recorded from Ireland. Analysis of the habitat data suggests a strong association with synanthropic habitats, especially cultivated sites, open spaces and churchyards. In Europe it is a subterranean species typically associated with deciduous woodland on base-rich, clay soils but is also found on light, acidic soils and in pine woodland (Kime, 2004). The species has been found in semi-natural woodlands in Britain (25% of records of the species) and almost all of these records have been from calcareous soils. However, due to the relative scarcity of the species, the analysis does not suggest a strong association despite 79% of records being from calcareous soils. The British data give no indication of a preference for clay soils. It is a common, central European species recorded from most countries north of the Alps. Kime (1999) suggests the species is probably at the limits of its range in Britain which may explain its synanthropic tendencies as well as its absence from Ireland and its scarcity in the west. Both males and females survive for several years after maturing (Blower & Miller, 1977) hence adults may be found throughout the year. However its habit of burrowing in adverse conditions may make it difficult to find in especially cold or dry conditions.

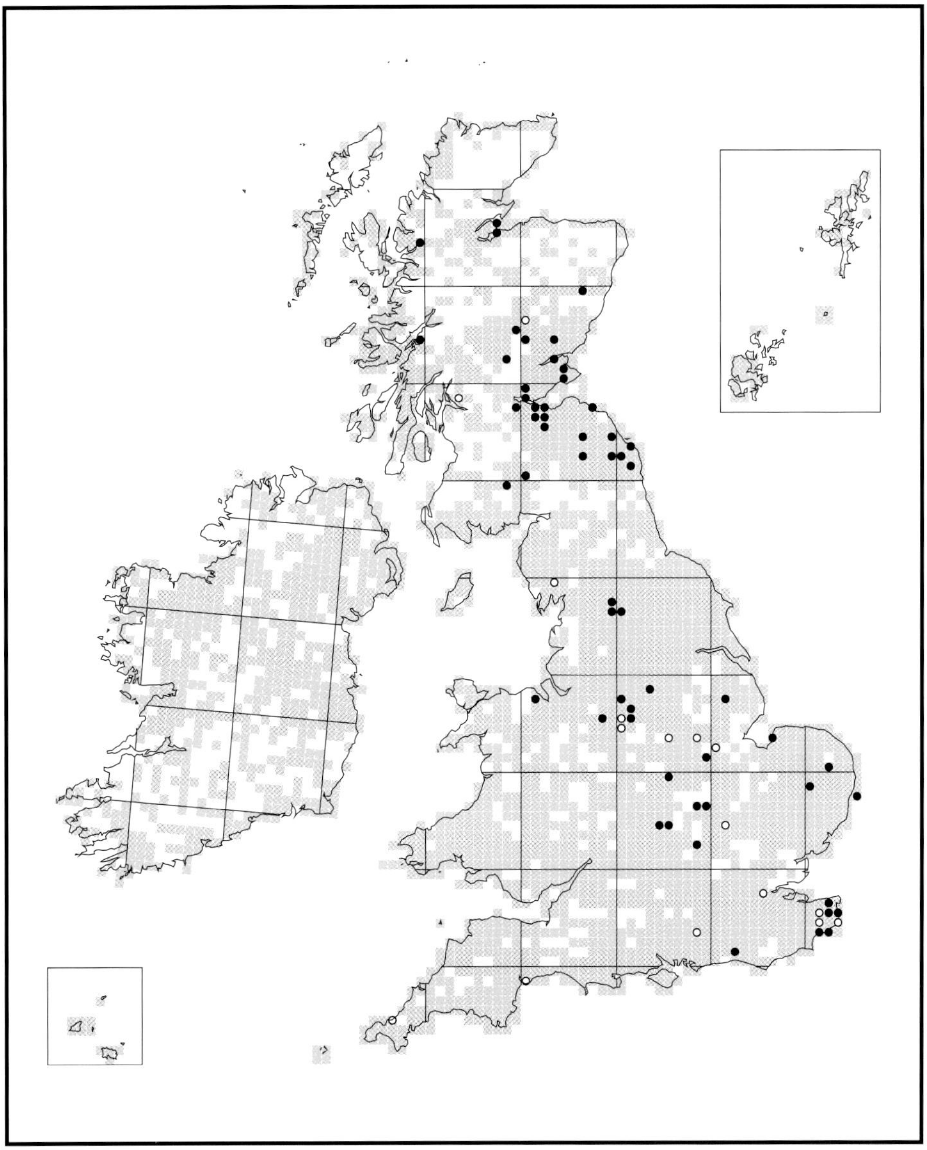

Cylindroiulus britannicus

Adult specimens, ideally males, should be dissected in order to differentiate populations of this species from *C. latestriatus.* Thus, although *C. britannicus* is a widespread species in Britain and Ireland, records are clustered around the locations of active myriapodologists capable of distinguishing the species. Kime (1999) described the species as "wild" in Britain and Ireland but synanthropic in continental Europe. Although 69% of the British and Irish records are from rural sites, analysis of the data suggests a strong relationship with urban sites as elsewhere in Europe. Although this species has been collected from purely synanthropic sites such as compost heaps and glasshouses, it is more usually collected from beneath the bark of standing and fallen dead wood or in the soil close to fallen timber, even in urban locations. The species appears to be most frequent in Britain and Ireland but is known also from widely scattered locations across northern Europe. There are two Portuguese records (Kime, 1990a) and specimens have recently been collected in north west Spain which, although it has yet to be found in western France, may suggest the species originally had a strict Atlantic distribution. It has also been introduced to New Zealand, North America and South Africa (Blower, 1985). Animals probably mature after two or three years as with *C. latestriatus*. Adults have been found throughout the year. This raises the possibility that, as with *C. latestriatus,* females may survive beyond their first breeding season.

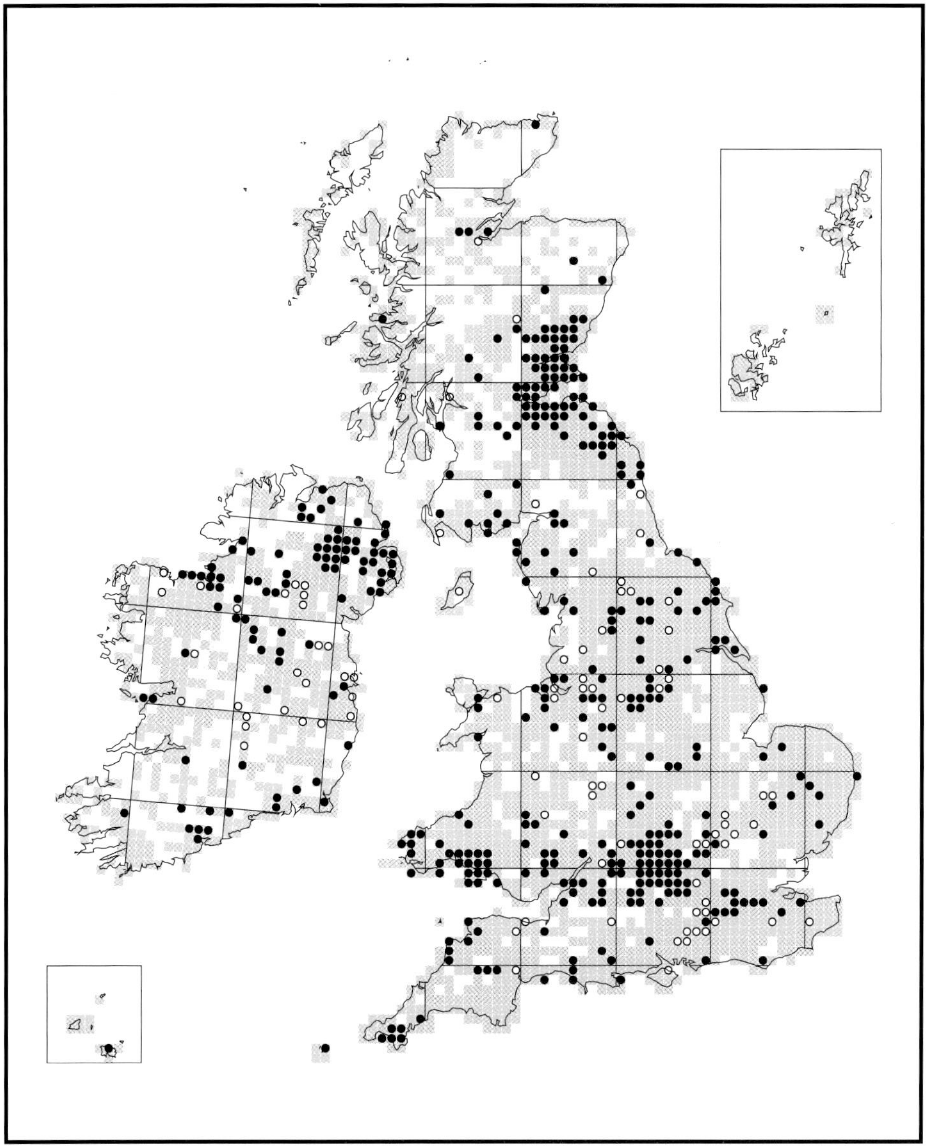

Cylindroiulus caeruleocinctus

This species is found most frequently in the south and east of England. There are only scattered records from Wales and from south west England but its range extends northwards as far as Aberdeen along the east coast of Scotland. The few Irish records are mainly restricted to the southern and eastern coasts. Although this species has been collected from a wide range of habitats, analysis of the habitat data suggests it is synanthropic showing very strong associations with cultivated land and churchyards and a strong association with parks and other open spaces. In continental Europe the species is also synanthropic (Kime, 2004; Pedroli-Christen, 1993) but Kime (1997) found it to be common on grassland in Belgium, especially on calcareous soils. Blower (1985), Haacker (1968) and Pedroli-Christen (1993) have all noted a link with calcareous soils and analysis of the British and Irish data strongly supports such an association. However, Kime (1995) has suggested this association results from the millipede's ability to withstand the dry conditions on these soils rather than a requirement for a base-rich substrate. This species has an extended Atlantic distribution from Portugal to southern Scandinavia and extending eastwards through the Baltic. It is most frequently recorded in a restricted area stretching from the Alps to Denmark and southern Sweden (Kime, 1999). Animals move deep into the soil to overwinter or to aestivate during dry summer conditions. Peaks of activity occur as they emerge in the spring and autumn (Kime, 2004). Adults have been found throughout the year.

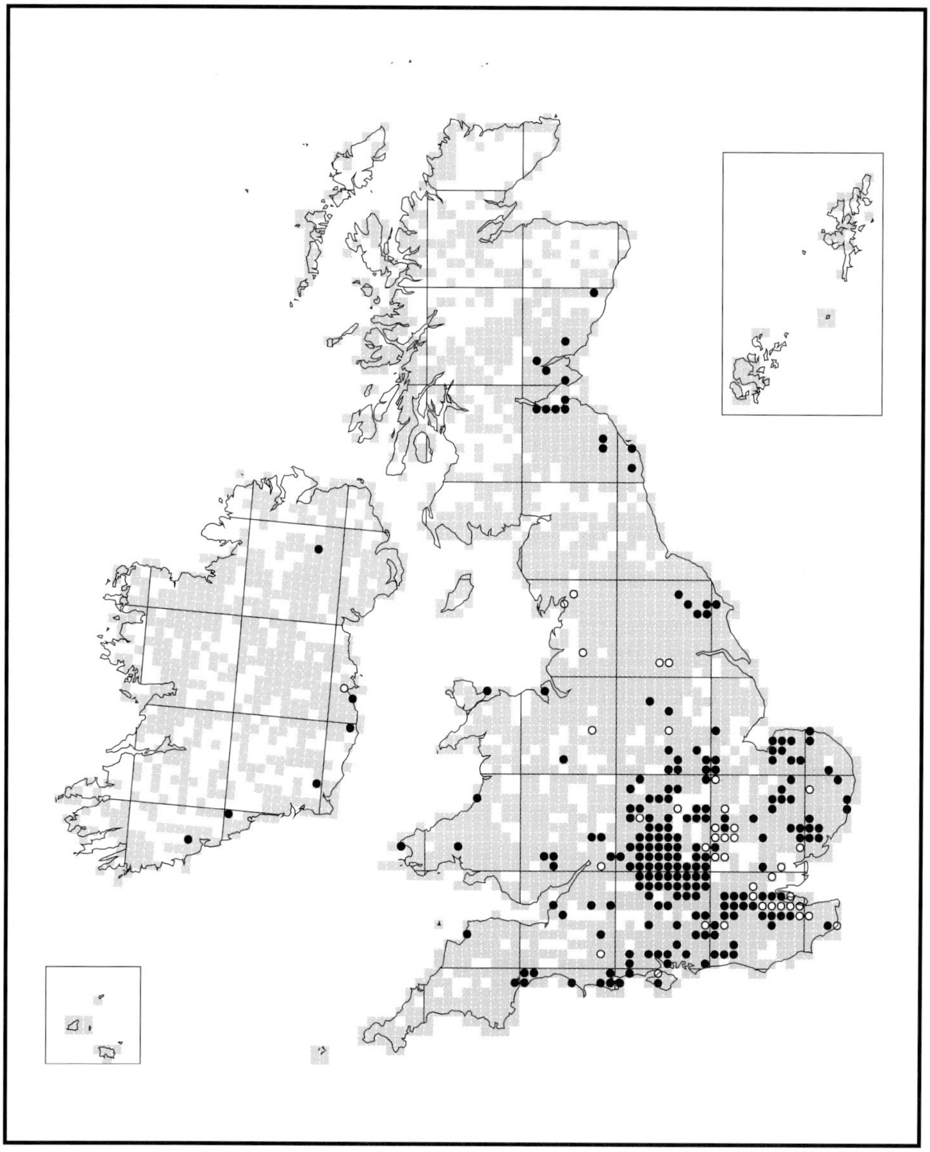

Cylindroiulus latestriatus

A widespread and common species all around the coasts of Britain and Ireland, this animal is sometimes encountered inland, most commonly in the East Anglian brecklands. Analysis of the habitat data suggests a very strong association with maritime habitats and especially with dune systems. The species is very strongly associated with sandy soils hence its occurrence in Breckland. Although not evident from the habitat analysis, many of the other inland records relate to synanthropic populations. Elsewhere in Europe the species shows a similar marked association with sand. It mainly occurs in dune systems and coastal grassland along the Atlantic and North Sea coasts but in Belgium has been found in scrub and woodland as well as in arable land (Kime, 2004). It has an Atlantic distribution from Portugal to Norway and as a synanthrope it occurs eastwards to Russia (Kime, 1999). Outside of Europe it has been introduced to places as remote as Easter Island and Antarctica (Blower, 1985). Adult females are present throughout the year as, although males die after reproduction, the females survive for three or more years after reaching maturity (Blower, 1985). Blower & Gabbutt (1964) found that animals in a Devon woodland took two years to reach maturity but Desmukh (in Cotton & Miller, 1974) found that animals on sand dunes in Scotland required three years. Adult specimens, ideally males, should be dissected in order to differentiate populations of this species from *C. britannicus*.

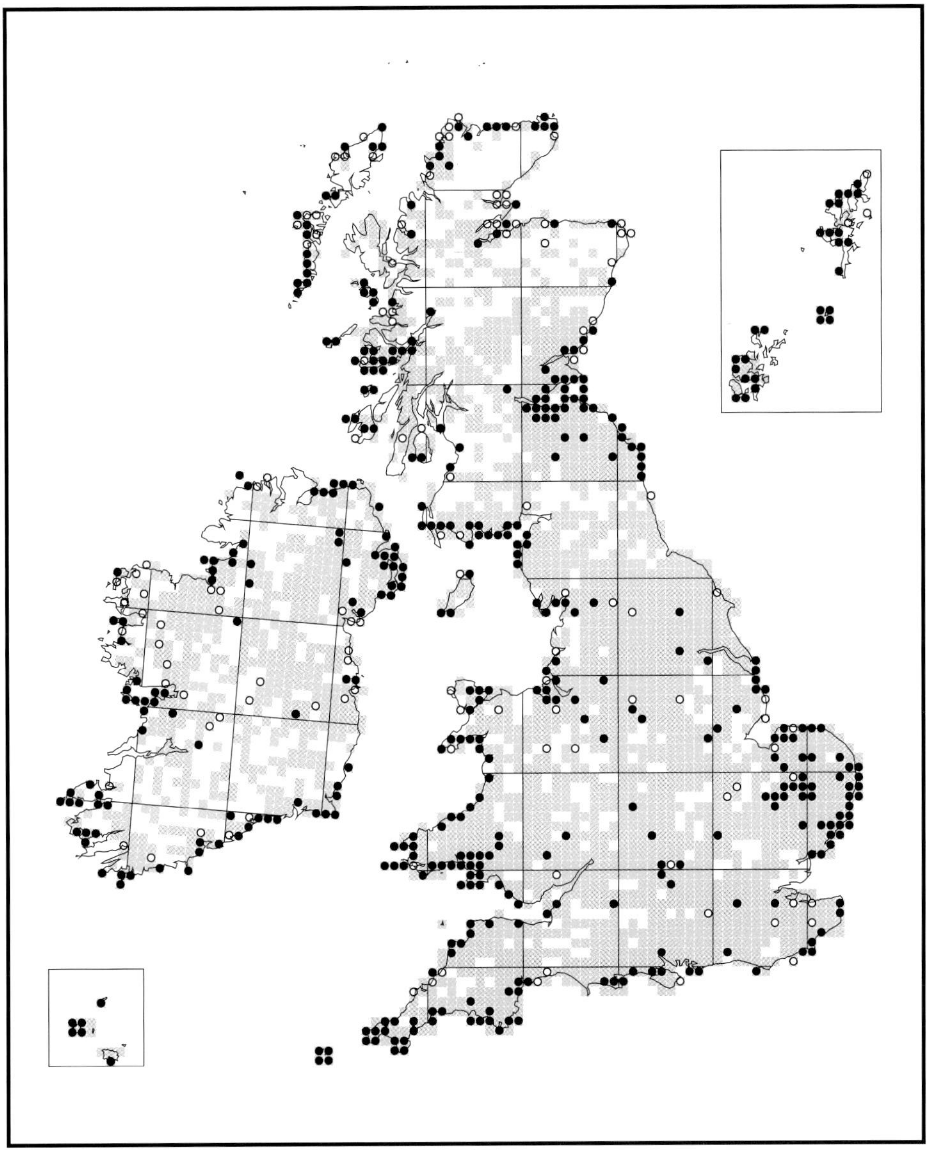

Cylindroiulus londinensis

This species is widespread but uncommon in Britain and Ireland. It has been recorded most frequently from south east England with few records from Scotland, Wales or Ireland. In Britain and Ireland, it has been recorded most frequently from woodland (48% of records) but the analysis does not suggest an association with this habitat. It has been recorded in a range of different habitats but not so far in heathland, moorland or wetland and not in strongly synanthropic sites. In Europe, Kime (1990a) considered *C. londinensis* to be a woodland animal whilst Demange (1981) gave it as frequent in woodland and open land up to 2 000 m. Kime (1999) considered the species to be most frequent on calcareous soils but noted it also occurred on free draining acidic soils. Analysis of the habitat data does not support an association with calcareous soils although there is some evidence for an association with free draining sandy soils. *Cylindroiulus londinensis* has a strict Atlantic distribution. In addition to Britain and Ireland, it is only known from the western half of France and from northern Spain (Kime, 1999). Adults have been found in each month from April to August and in November and December.

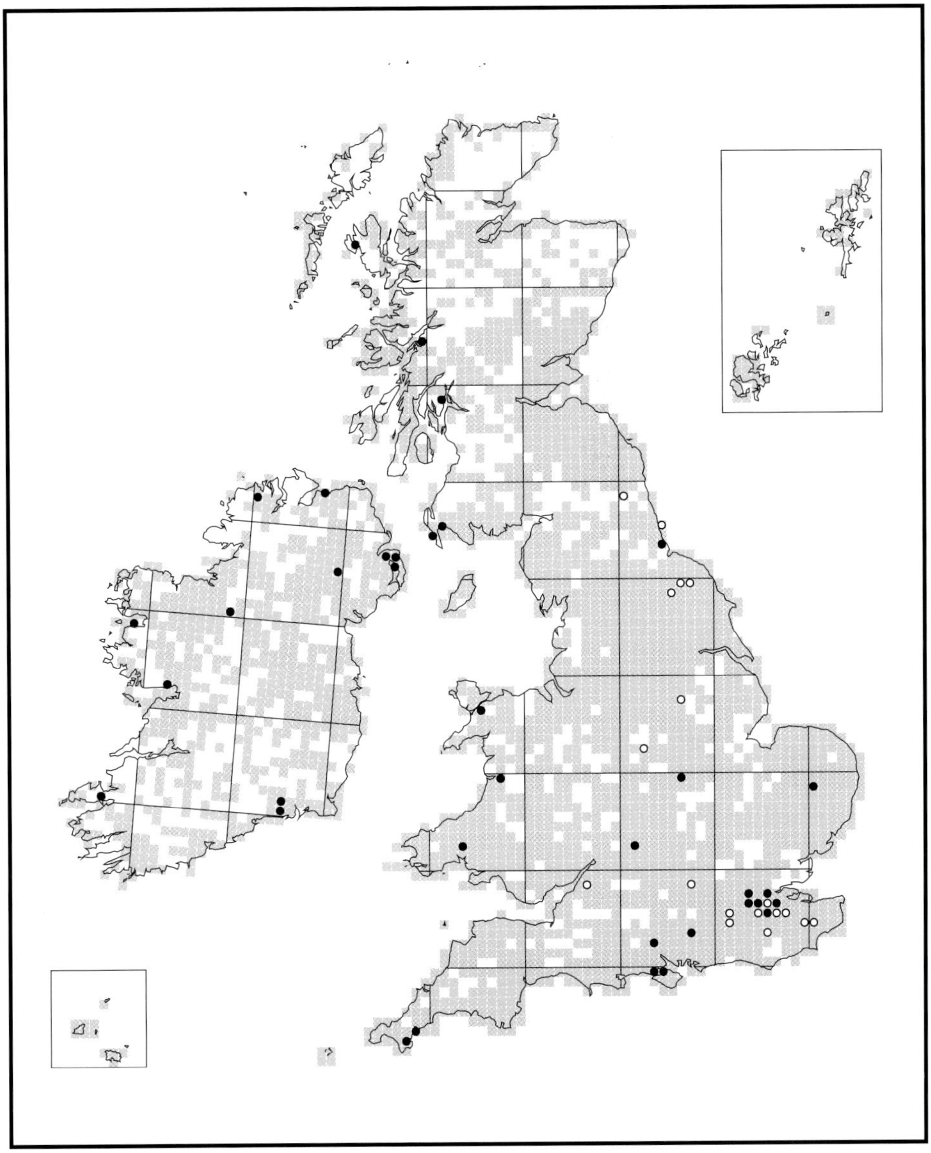

Cylindroiulus parisiorum

The distribution of this species appears to be centred on East Anglia and southern England. It has been recorded just twice from Wales, once from Ireland and not at all from Scotland. In Britain the species has been recorded from synanthropic sites as well as semi-natural grassland and woodland. The only Irish record was from a synanthropic population (Doogue *et al.,* 1993). Analysis of the habitat data suggests no strong association with any of the habitats or environmental factors considered. In continental Europe it has mainly been recorded from synanthropic sites but also from semi-natural habitats in Belgium (Kime, 1999) and in Switzerland (Pedroli-Christen, 1993). In semi-natural habitats it is usually found under loose bark or otherwise closely associated with decaying wood (Kime, 2004). Many of the British records from synanthropic sites are also associated with decaying wood. The species has been most frequently recorded in Britain but is known also from widely scattered locations across central Europe (Kime, 1990a). Confusion with *C. truncorum* may explain, at least in part, the dearth of central European records (Kime, 1999). Adults have been collected in each month from February through to June and September through to December. It remains to be seen whether adult females are also present in July and August and hence survive to breed in more than one season.

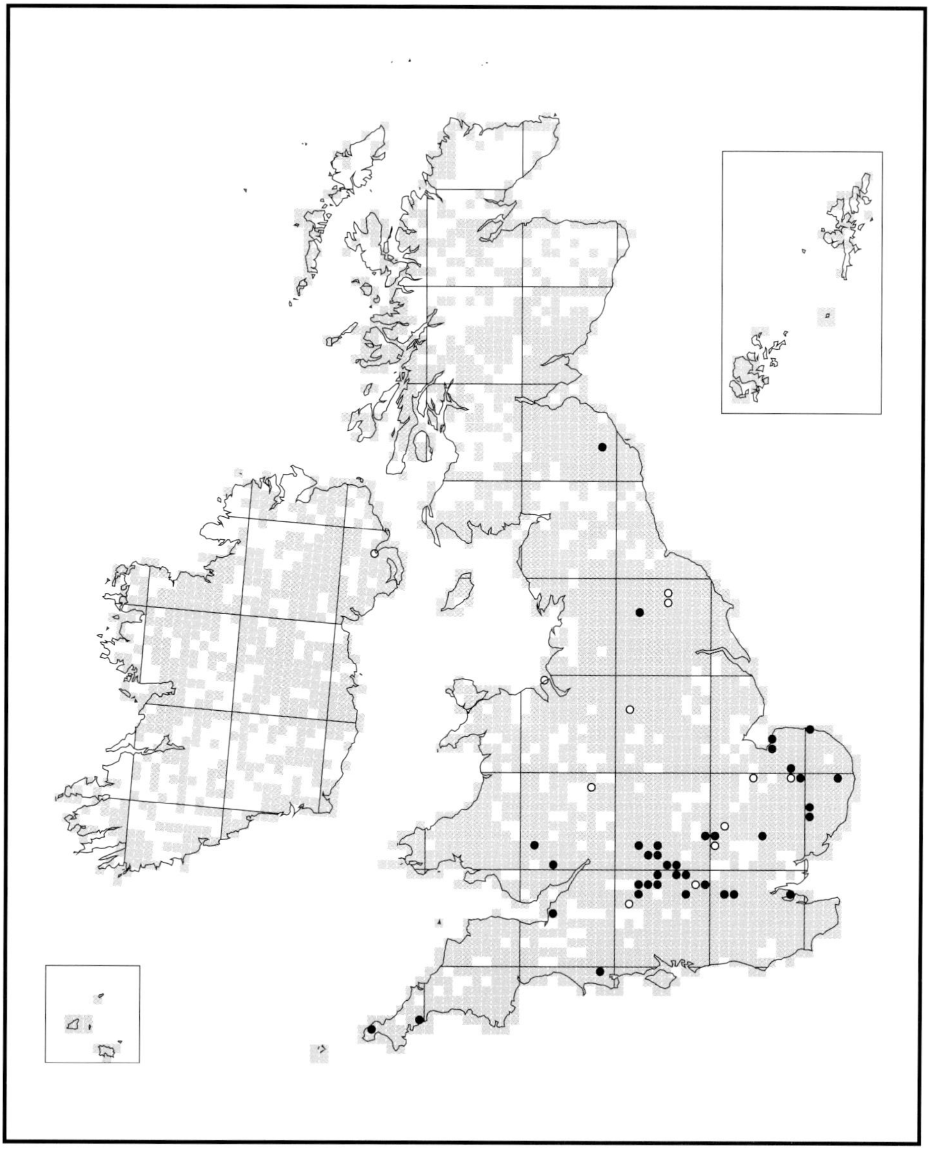

Cylindroiulus punctatus

This is the most common and widespread species in Britain and Ireland. It is found in most habitats but analysis of the habitat data suggests a very strong association with woodland and strong negative relationships with more open grassland and sand dune habitats. It is an animal of rural, semi-natural habitats rather than synanthropic sites. Elsewhere in Europe it is also typically a woodland animal (Kime, 1990a). The habitat analysis also suggests a strong association with non-calcareous soils; something not found by Kime *et al.* (1992) but they did note a preference for loamy soils. There is some evidence (51% of records) of a preference for loams in Britain and Ireland also, even though the habitat analysis does not suggest a strong association. Banerjee (1967) described the seasonal migrations of *C. punctatus* brought about by changing temperatures. Animals move from leaf litter into decaying wood in the spring and then return to the litter in autumn, ultimately overwintering in the mineral soil if conditions are sufficiently cold. *Cylindroiulus punctatus* has an Atlantic distribution from northern Spain to southern Scandinavia. Temperature may be a key limiting factor. Low winter temperatures probably limit its northern and eastern distribution. In southern France it becomes much harder to find, possibly due to high spring and summer temperatures (Kime, 1999). Blower & Gabbutt (1964) found that animals in a Devon woodland took three years to mature and females may survive and breed for several years beyond that. This means adult females may be collected throughout the year.

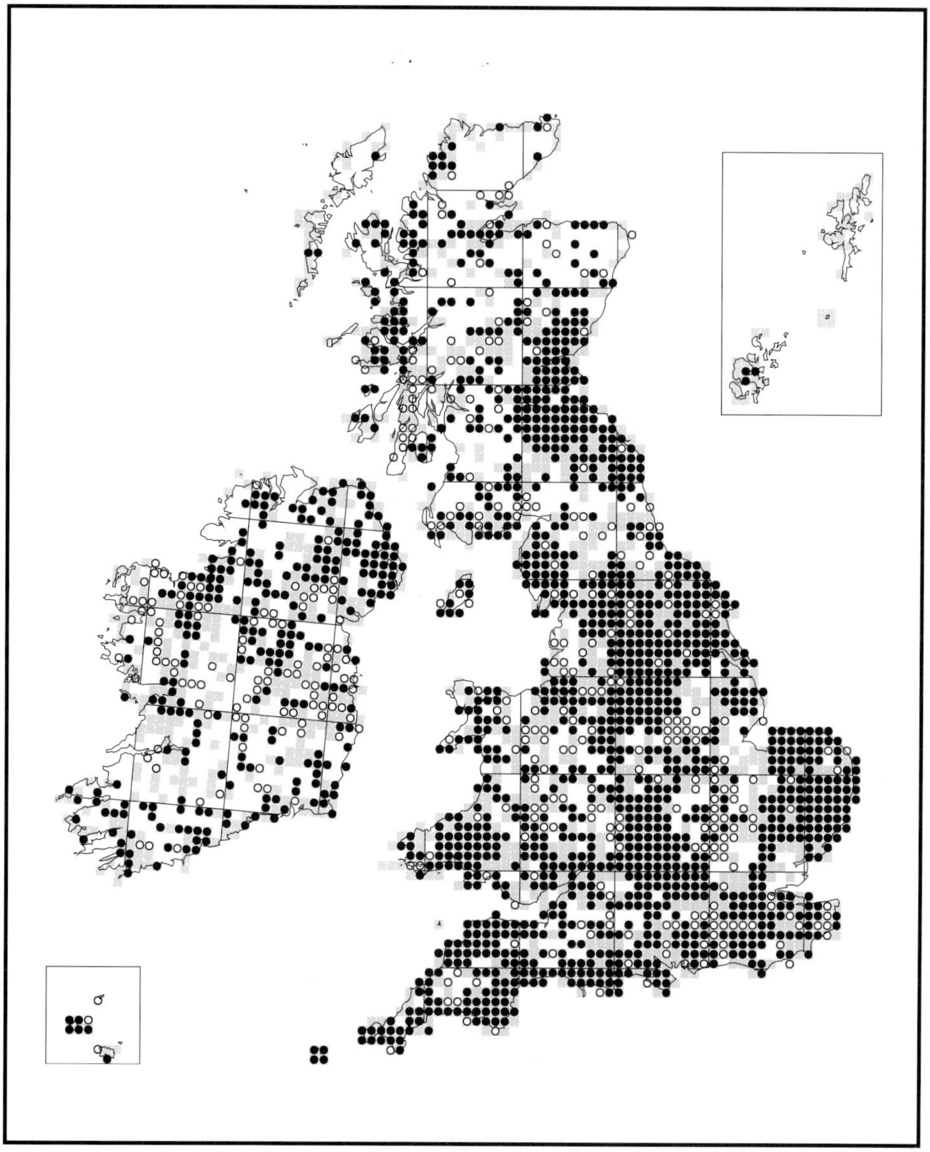

Cylindroiulus truncorum

This species has been recorded from the botanic gardens at Kew, Oxford and Glasgow and from a site in Leicestershire. The first British record from "near Bristol" (Lindroth, 1957) is too imprecise to be mapped. Most recently it was collected from Mount Stewart, Northern Ireland and several outdoor sites on Jersey. In Britain it has been collected from both inside hothouses and from leaf litter outdoors but there are insufficient ecological data available for reliable conclusions to be drawn from any analysis. Kime (2004) gave a range of synanthropic habitats for the species in Belgium including reclaimed spoil heaps. Haacker (1968) noted that it occurred in woodland in Germany and that it preferred high humidity. *Cylindroiulus truncorum* can be common where it occurs at scattered locations through northern Europe and may have been confused with *C. parisiorum* in the past. It appears to be absent from Iberia and Italy (Kime, 1999). Schubart (1934) suggested it was probably introduced into northern Europe from North Africa. It has also been introduced to North and South America (Schubart, 1946). The life history is probably similar to that of *C. britannicus* and *C. latestriatus* but with both males and females undergoing one more moult before reaching maturity than in those species (Blower, 1985). Adults have been collected in August, September and November.

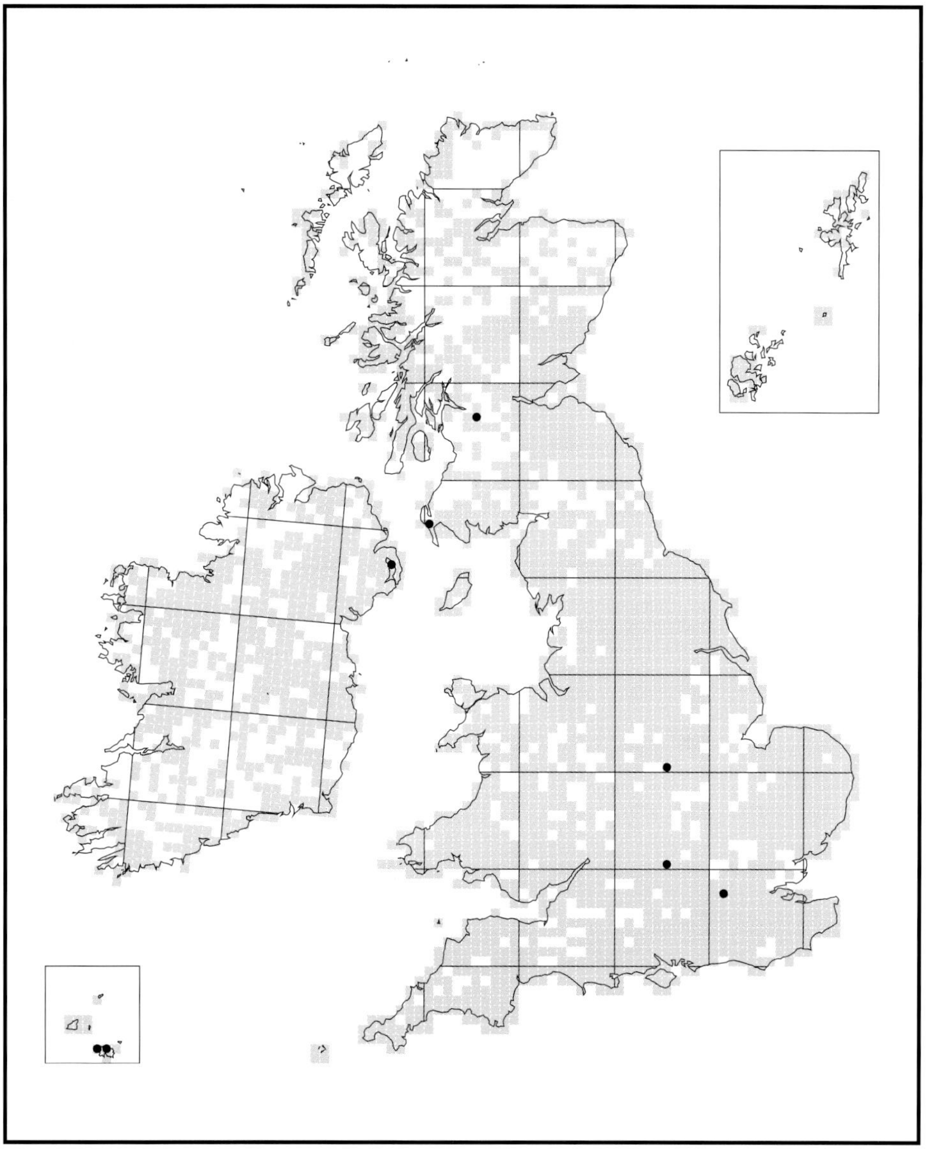

Cylindroiulus vulnerarius

This species was collected in Britain for the first time in 1975 and in Ireland in 1978. Since then it has been recorded occasionally from approximately twenty, mainly urban sites across Britain and Ireland. As Gregory & Campbell (1996) reported from Oxford, the millipede may be abundant where it is found. Analysis of the habitat data suggests that this species is strongly synanthropic. It is particularly associated with glasshouses, gardens, parks and other open spaces in urban areas. Kime (2004) considered its habitat to be rotting wood and other decaying organic matter in contact with the soil. *C. vulnerarius* is supposedly endemic to Italy. It has been widely introduced into synanthropic habitats in northern Europe, especially in Belgium and the Netherlands but also in France and Sweden (Kime, 1999). However, its recent discovery in a cave raises the possibility it may be a relict species in Belgium (Kime, 2004). Bilton (1990) monitored populations of this species at two sites in Oxford and found it active close to the soil surface between November and April only. He also found animals kept in captivity were surface active at night. Adult animals have been collected from October through to June. The apparent absence of adults in the summer may indicate that both sexes die after reproduction but as the total number of records of this species is so small it is still possible that adults will be found in July and August.

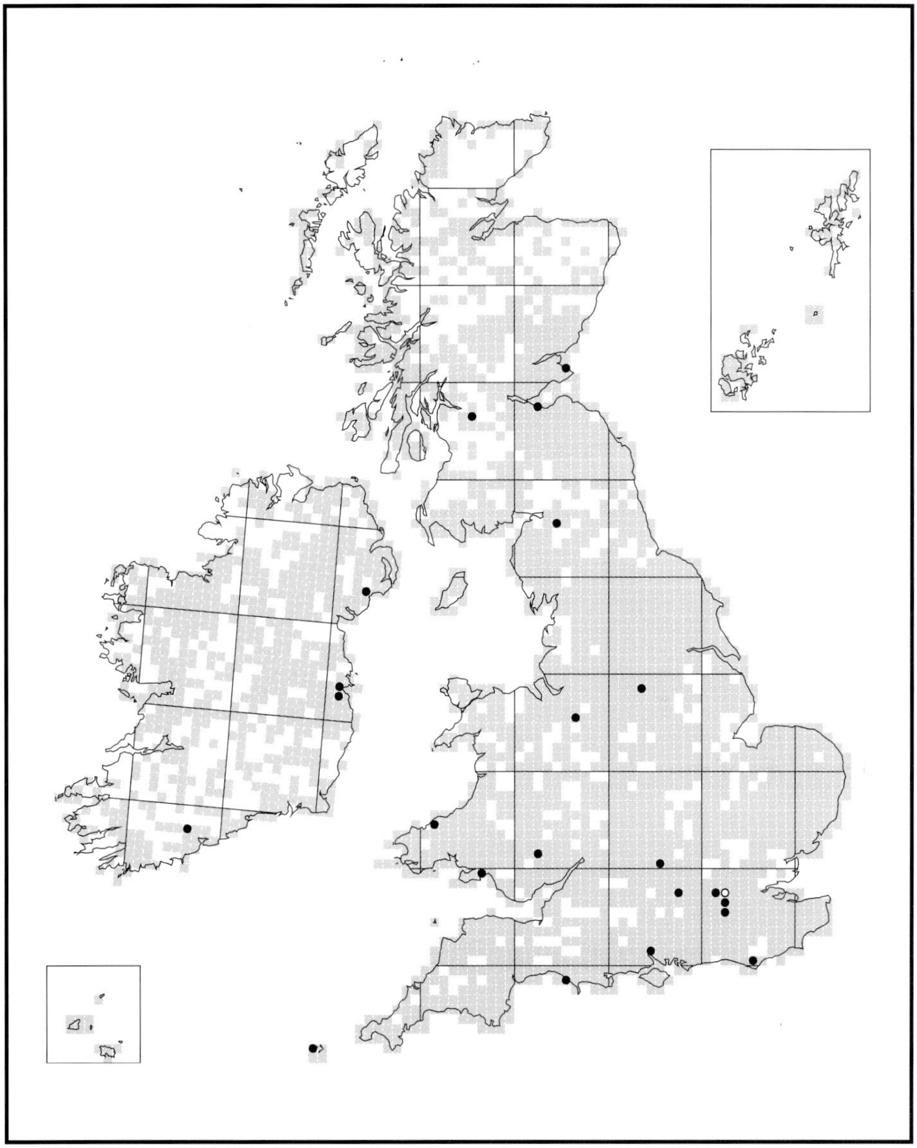

Enantiulus armatus

This species is most frequently recorded from south Devon but has also been found in Cornwall and north Devon. It was originally collected in 1958 from leaf litter in deciduous woodland and from a golf course, both near Sidmouth (Blower, 1985). Subsequent records have been from a range of coastal, synanthropic and woodland sites but the latter include plantations of non-native trees (Barber, 1987; Bolton, 1996). Analysis of the habitat data suggests no strong association with any of the habitats or environmental factors considered. *Enantiulus armatus* has a strict Atlantic distribution and outside of south west England it is known only from south west France (Kime, 1999). The presence of the species in synanthropic habitats may suggest it has been introduced to Britain. However, it is also found within a variety of habitats in France (Kime, 1990a) so there is no reason at this point in time to assume it is anything other than native. The species has been recorded in most months of the year but there are confirmed records of adults from February and May only.

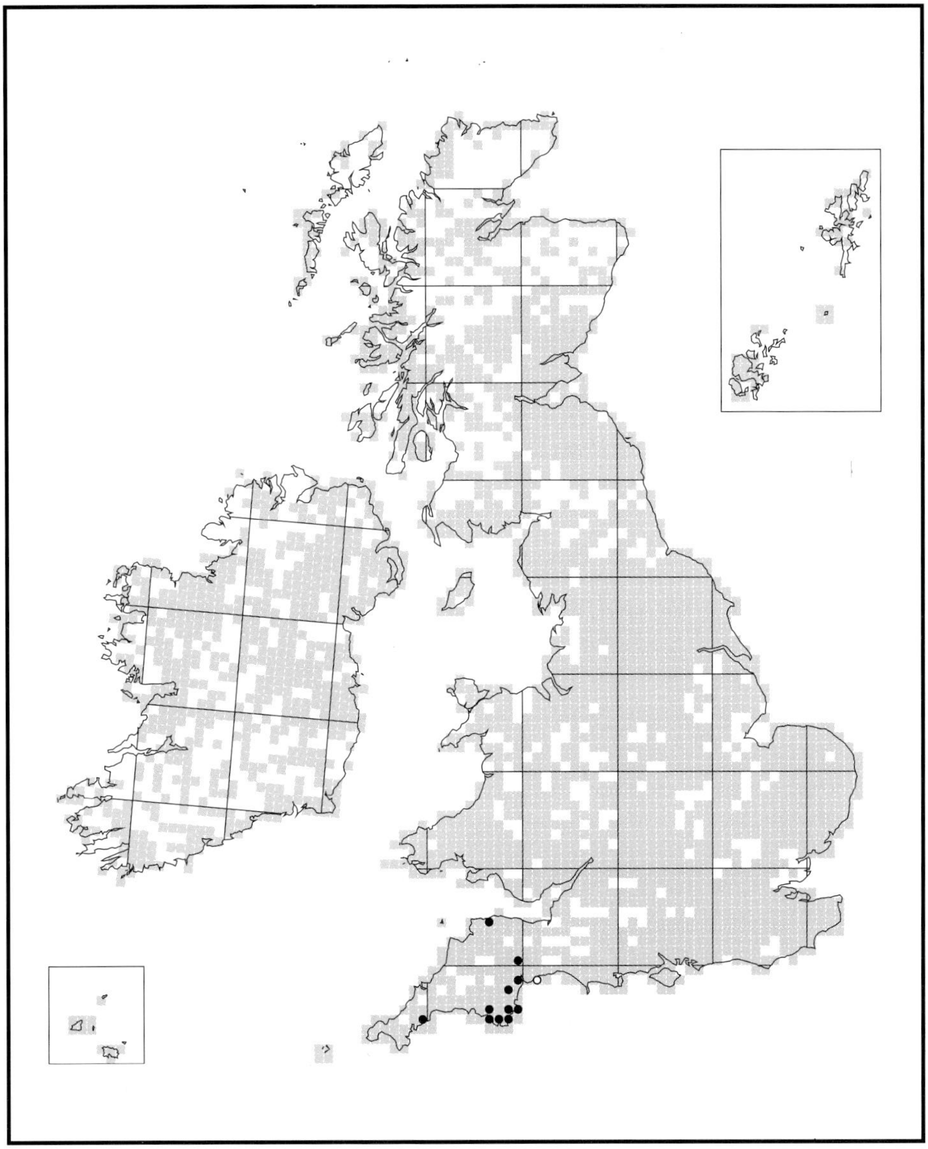

Unciger foetidus

This species was almost certainly introduced to its only known British location, a garden in Dersingham, Norfolk, where it was first recorded in 1983. It has been reported from synanthropic habitats elsewhere in north west Europe including Denmark (Enghoff, 1974) and Norway (Meidell, 1968). Blower (1985) mentions its occurrence in the Netherlands but this record is no longer considered valid (Berg, 1995; Jeekel, 2001) and it has not been recorded from France or Belgium. *Unciger foetidus* appears to be native to woodland in central Europe, occurring as far east as Ukraine. Haacker (1970) found it required high humidity consistent with it being a woodland species. It is also thermophilic but tolerates a wide range of soil acidity. It becomes increasingly synanthropic towards the north and west of its range (Kime, 1999) and it could conceivably be discovered in gardens elsewhere in Britain and Ireland. Adults have been collected in April in Norfolk.

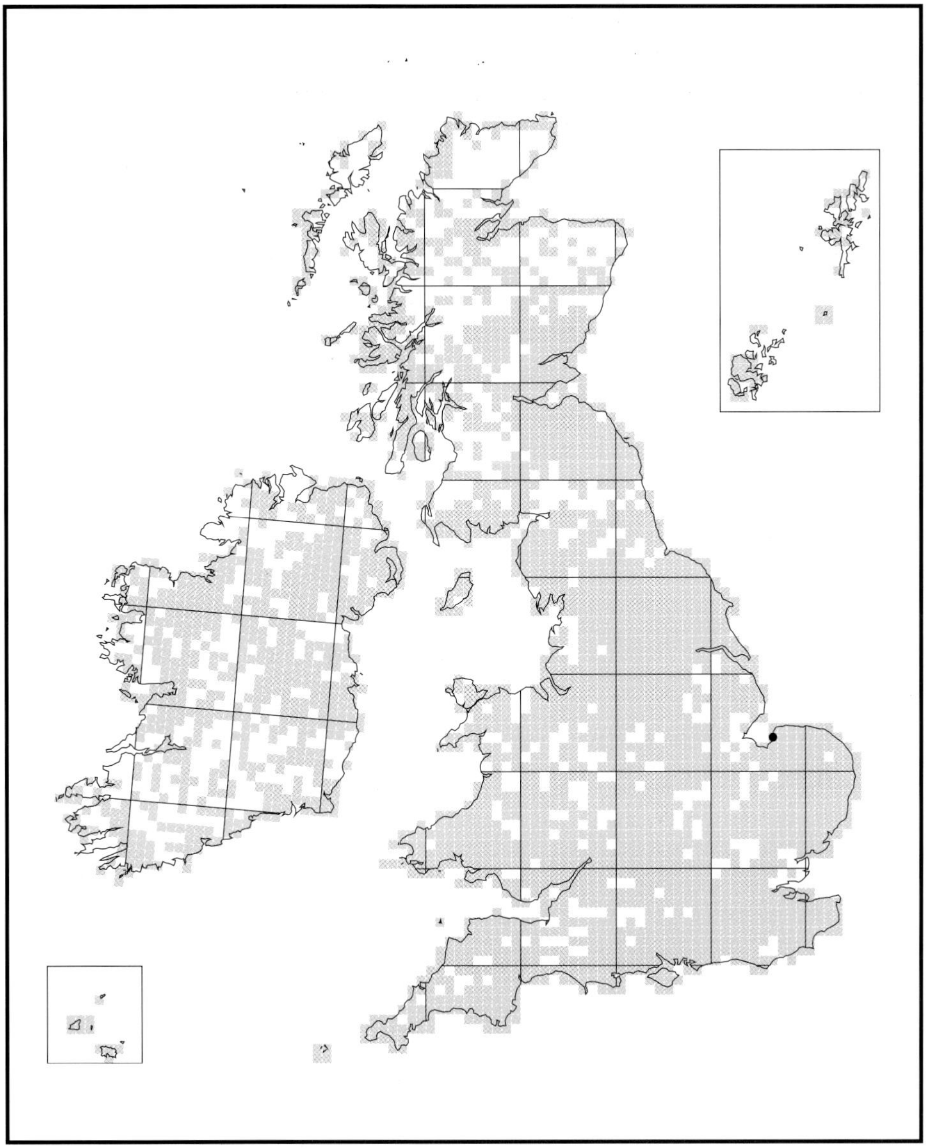

Brachyiulus pusillus

This species is widespread across Britain and Ireland. Blower (1985) shows records for vice counties 96, 97, 98 and 104 in NW Scotland. These were listed but not mapped in the preliminary atlas (British Myriapod Group, 1988) and are not mapped here as precise grid references were not available. The species appears to become increasingly coastal further north in Britain but also shows a strong association with cultivated sites and grasslands at inland locations in the south. Gregory and Campbell (1996) noted it was typical of river-side meadows in Oxfordshire. Analysis of the habitat data suggests a strong negative relationship with woodland, something which Kime (2004) has noted in Belgium along with an association with pastures and agricultural land. Analysis of the soil data suggests a strong association with clays but a negative relationship with loams. Jeekel (2001) linked the species with humid, clayey soils in the Netherlands and suggested it could survive periodic inundation. *Brachyiulus pusillus* is also widespread in Europe, although Kime (2004) noted that some of the eastern records may refer to another species, *B. bagnalli* Brölemann, 1924. *Brachyiulus pusillus* has been introduced to North and South America, South Africa and many of the Atlantic islands (Blower, 1985). Adults have been collected throughout the year but records are most frequent in the spring between March and June. Both sexes probably die after breeding (Blower, 1985) in the spring.

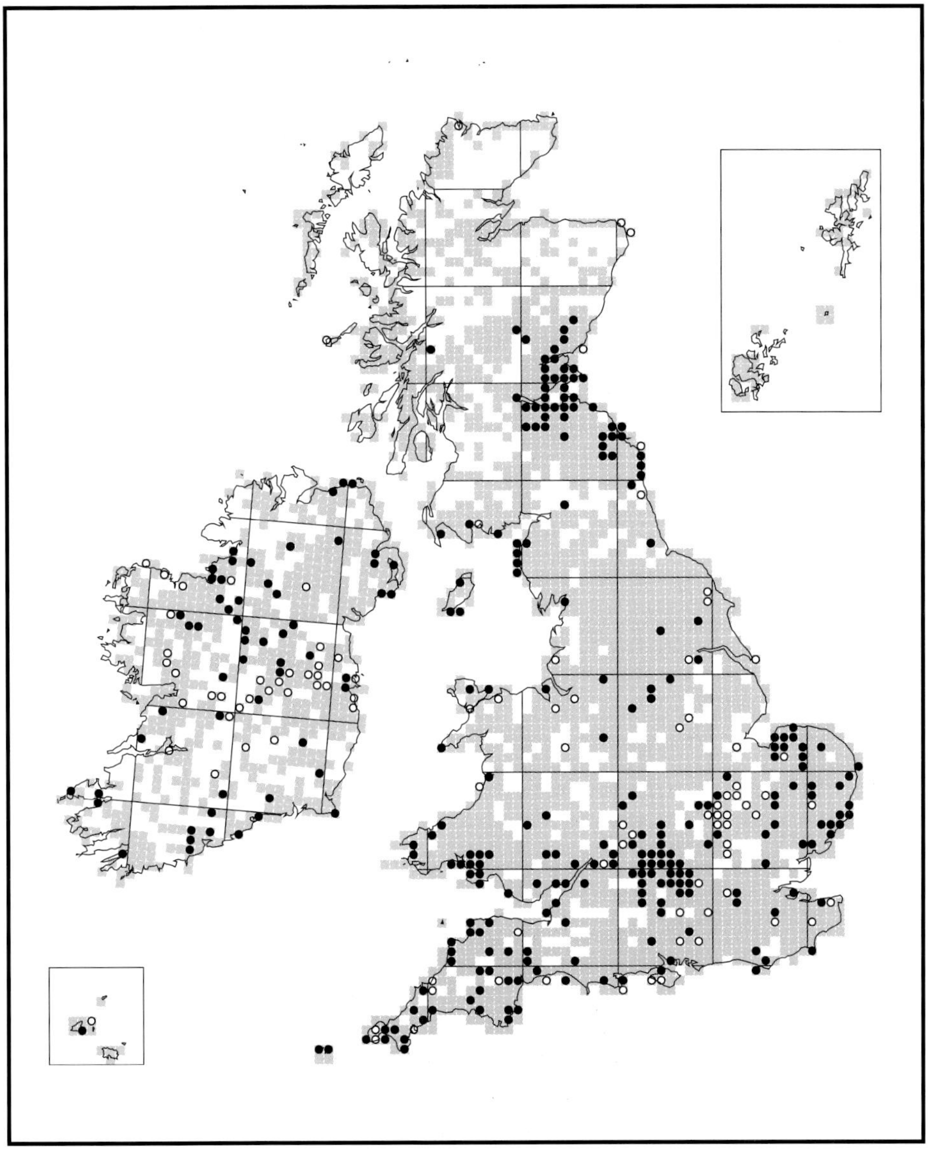

Ommatoiulus sabulosus

Although generally widespread and common in Britain and Ireland, this species is surprisingly scarce in central southern and south eastern England. Adults wander widely in the summer and can occur in almost any type of habitat. However, analysis of the data suggests this is largely a coastal species in Britain and Ireland showing a very strong association with sand dunes and a strong association with other maritime habitats and with heathland. A very strong association with sandy soils clearly links these habitats. Barlow (1957) and Biernaux (1969) also found that, in the Low Countries, the species preferred warm, well drained soils. However, the suggestion of a strong association with peaty soils in wetland and moorland habitats is surprising and not supported by work elsewhere in Europe. This is one of the most widely distributed millipedes in Europe although it is absent from Iberia, Greece and many Mediterranean Islands (Kime, 1999). Adults can be found throughout the year but they are less frequently recorded in winter. At this time of year they appear to be less resistant to desiccation than in the summer and so are not actively wandering (Kime, 2004). In winter the adults can be found by searching through moist layers of soil and leaf litter. Fairhurst (1974) reported that males and females from a population living in sand dunes on Anglesey matured after two or three years respectively and one year earlier than the equivalent sex living in adjacent conifer woods.

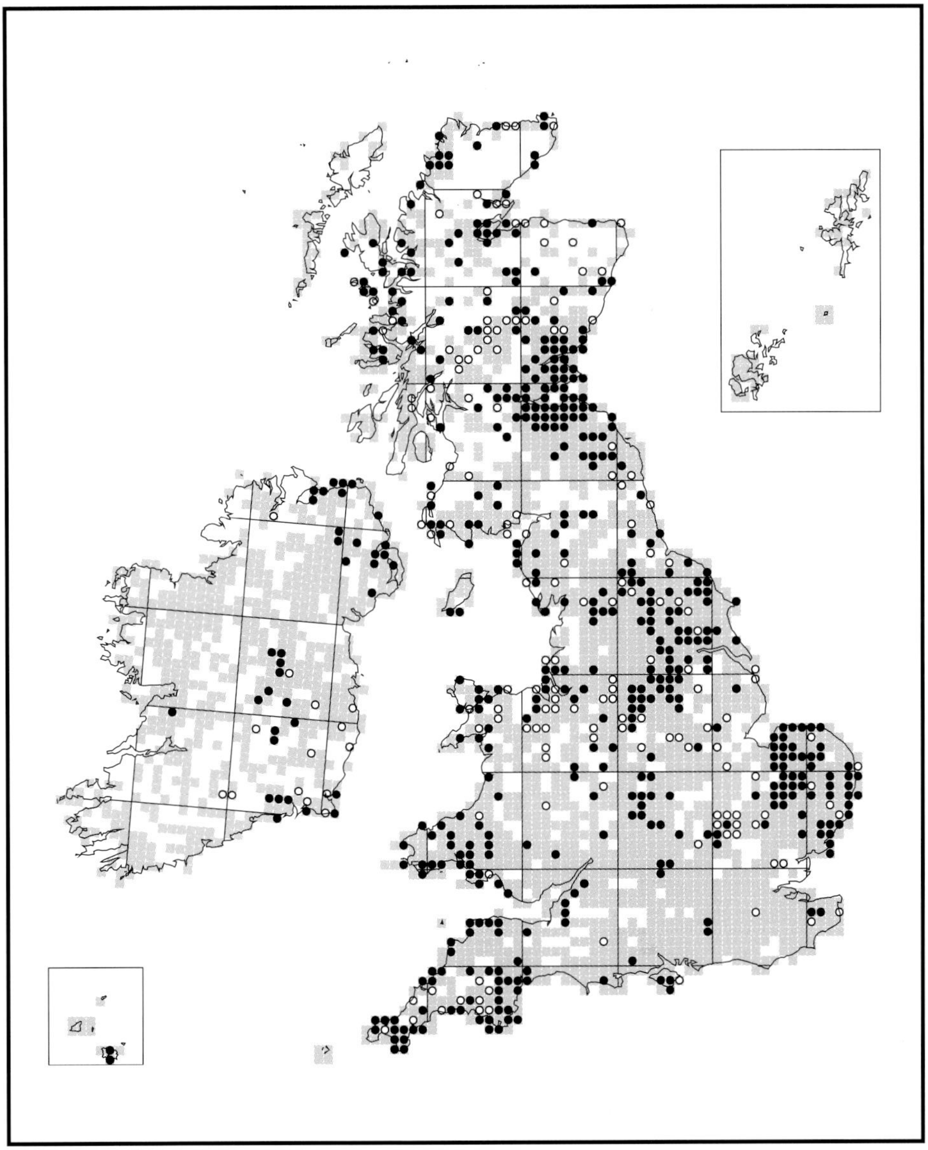

Tachypodoiulus niger

This is one of the most commonly recorded millipedes in Britain and Ireland. It has a very widespread distribution and can attain high population densities in suitable habitat but its high levels of activity account for the frequency with which it is encountered by myriapodologists and the general public alike. Analysis of the habitat data suggests this is an inland rather than a coastal species in contrast to *Ommatoiulus sabulosus*. A strong association with waste ground is suggested by the analysis but this is another very active animal which could be encountered in almost any habitat. Kime (2004) reported *T. niger* to be strongly associated with woodland in Belgium. Although 43% of the British and Irish records were from woodland, analysis of the data did not support the existence of a strong association here. Kime *et al.* (1992) found that the species was most abundant on limestone but analysis of the recording scheme data provides no evidence of a preference for calcareous soils. This species has an extended Atlantic distribution in continental Europe where it is found from the Pyrenees north to Germany and east to the Czech Republic (Kime, 1999). Adults can be found throughout the year but are less frequently collected in winter. Animals mature after two or three years (Blower & Fairhurst, 1968) and may survive for several years after first mating. The largest females may be up to nine years old (Blower, 1985). Adult animals readily climb trees and are often collected by entomologists using beating trays.

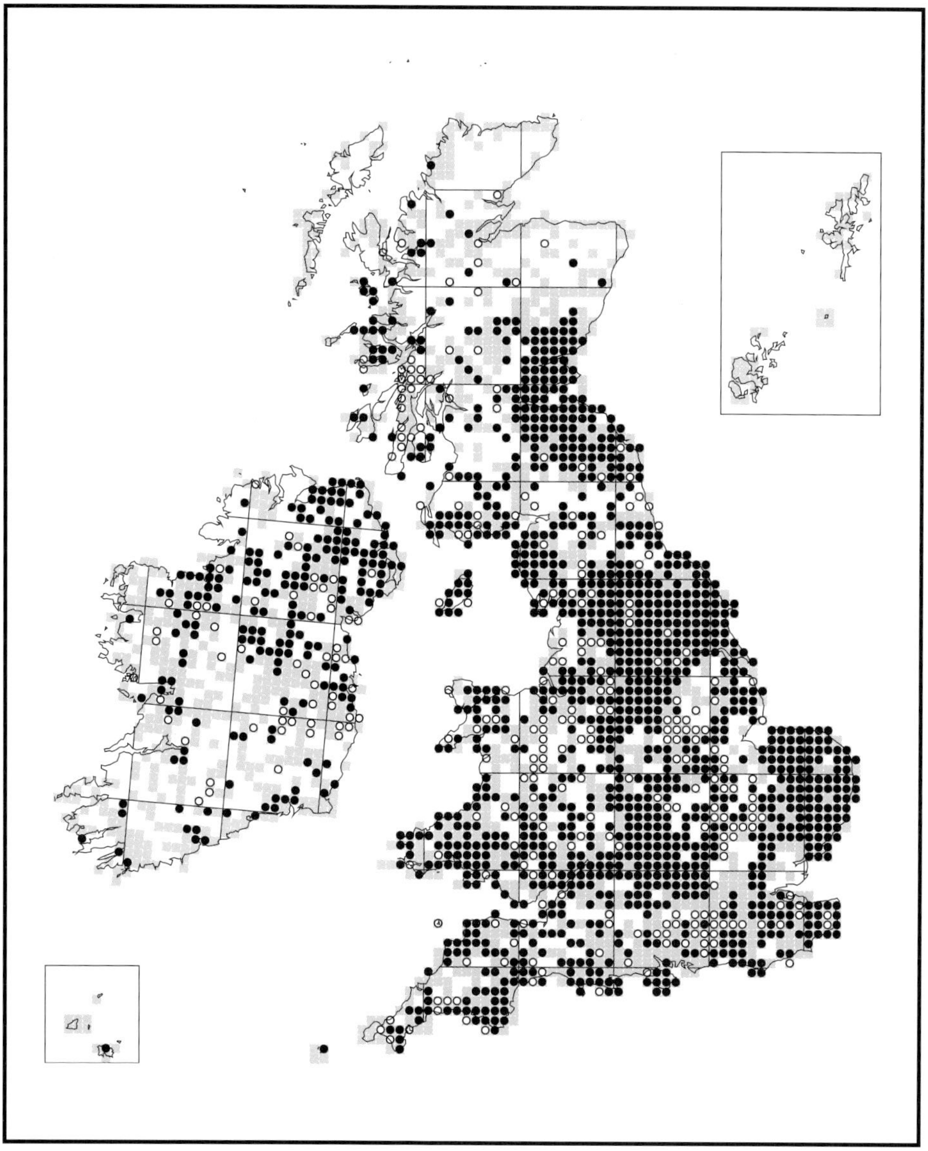

Alien Species

Included here are six species which have been introduced to Britain and Ireland but, unlike many of the introductions, have not established themselves out of doors. These millipedes occur mainly within heated glasshouses where they have established viable populations. Casual introductions which have not established breeding populations and alien species which have not been collected since before 1980 are not considered here. A brief account of each species is given along similar lines to the other species accounts. The distribution of these species has not been mapped but is summarised at the end of the account as the 10km squares in which it has been recorded post-1980.

Oxidus gracilis

At the present time this species appears unable to survive outdoors in Britain and Ireland. However, it is well established in nurseries and botanic gardens across Britain and is also known from two locations in Ireland. It was probably originally a native of East Asia (Blower, 1985) but is now found throughout the tropics where it survives outdoors and in many temperate regions, including the whole of Europe, where it may be found in glasshouses, gardens and adjacent compost heaps. Causey (1943) reported that clutches of eggs may be laid throughout the year and animals may take as little as five months to reach maturity at a temperature of 22° C. In addition to the established populations it is often recorded as a vagrant in pot plants in garden centres and commercial and domestic premises. Recorded from OS grid squares: SX05, SS69, SU57, TQ17, SN58, SP00, SP50, SK33, SK58, SE23, NZ24, NT27. Recorded from Irish grid square J37.

Cylindrodesmus hirsutus

Henrik Enghoff recalls seeing specimens of this parthenogenetic species from Britain around 1985 but the record was never published (Golovatch *et al.,* 2000). The first confirmed record is of individuals found in the Palm House at Kew in 2002. However, the identity of these animals was not recognised until further specimens were collected from the same location two years later. Further specimens were found at Kew in 2005 and a population was discovered in a tropical butterfly house in SouthYorkshire later the same year (Lee, 2005a). It is a pantropical animal and introduced populations have been found in hothouses in Berlin, Paris and Vienna (Golovatch *et al.,* 2000). Recorded from OS grid squares: TQ17, SK58.

Prosopodesmus panporus

This species was described as new to science on the basis of material collected from Kew Gardens in 1975. It inhabits the soil and litter of the Palm House, Tropical Aroid House No. 1 and Tropical Fern House No. 2 at Kew (Blower & Rundle, 1980) and is only known from this one location; it is still not known from which country it was introduced to Kew. Recorded from OS grid square TQ17.

Poratia digitata

This species was first collected from peat in the Tropical Fern House at Kew in 1985 and was then found under bark and in soil in the Orchid House at Leicester Museum Botanic Gardens later the same year. It has been collected at Kew on a number of occasions over the last 20 years but there were no other records until 2005 when it was found in Durham University Botanic Gardens (Lee, 2005b) and in a tropical butterfly house near Sheffield (Lee, 2005a). It has also been found in botanic gardens in Denmark, Switzerland and Germany and in a greenhouse in Sweden (Blower & Rundle, 1986). The species appears to be native to tropical South America and the Caribbean. Recorded from OS grid squares: TQ17, SK50, SK58, NZ24.

Poratia digitata

Paraspirobolus lucifugus

This species was first collected from the Tropical Biome at the Eden Project, Cornwall in 2005 (Barber, 2005). It probably originates from the islands of the Indian Ocean such as the Seychelles and/or Mauritius and is known to have been transported to various regions including Brazil and hothouses in Denmark and Germany. Recorded from OS grid square SX05.

Paraspirobolus lucifugus

Cylindroiulus salicivorus

This species was first found in Britain at the Royal Botanic Gardens, Edinburgh in 1987. It has also been collected from St Andrews Botanic Gardens (Read, Corbet & Jones, 2002). This millipede is endemic to northern Italy where it was originally found in wet grassland (Verhoeff, 1908). It has been reported as an introduction from nowhere other than Scotland (Kime, 1999). Adults have been collected in March, May and November. Recorded from OS grid squares: NT27, NO51.

Data analysis

The Data Set

More than 500 individuals and organisations submitted over 47 000 records to the Millipede Recording Scheme in the period from its inception in 1970 up until the middle of 2003. Of these records, 37 233 included details of at least the principal or first order habitat in addition to the geographic information required for mapping. Approximately 12 000 of these records were supplied on RA13 cards and 22 000 on RA59 cards. The remaining records were entered onto GEN7 record cards or supplied in electronic format. Although the data set does include older records extracted from the literature, there has been no attempt to undertake a comprehensive literature survey. The inclusion of records from the nineteenth and early twentieth century is largely dependent on enthusiastic individuals having researched the faunal records relevant to their local areas.

The influence of enthusiastic individuals is also evident in the regional variations (regions as defined by Barber & Keay, 1988) in the intensity of modern field recording (Table 2). A small number of very active field workers are responsible for the large number of records received from the South East and the Yorkshire, Lancashire, Westmorland regions. The paucity of records from Ireland, Northern Scotland and, to a lesser extent, Wales is a common feature of many invertebrate recording schemes (e.g. Alexander, 2003; Johnson, 1993; Luff, 1998).

Table 2: Number of records of millipedes submitted from the regions of the British Isles (comprising Britain, Ireland and the Channel Islands)

Region	Vice-counties	Number of millipede records
South-east	10–24, 30	11 213
South-west	1-9, 33–34	5 431
Wales	35, 41–52	4 098
East Anglia and east Midlands	25–29, 31–32, 53–56	5 589
West and north Midlands	36–40, 57–58	3 193
Yorks., Lancs., Westmorland	59–65, 69	7 125
North England, south Scotland	66–68, 70, 72-84	4 148
Northern Scotland	85–112	2 829
Ireland	H1–H40	3 478
Isle of Man	71	188
Channel Islands	113	140
	TOTAL	**47 432**

It is inevitable that a number of the smaller, less easily sampled species are under represented in the data set. For example, small soil dwelling species such as *Geoglomeris* or *Brachychaeteuma* may be most readily found by specialist extraction techniques and/or careful searching in surface layers of soil and under stones in winter months when fieldwork is a less attractive activity for most naturalists. In addition a number of taxa require the dissection of specimens, often males, for the accurate determination of species. Not only does this reduce the number of records likely to be submitted by non-specialists, it also means that even specialists may not submit records if only female specimens have been collected.

A small number of post-2003 records of a few rare millipedes, where any new data would improve our knowledge of the species, were included in the data set for the purposes of distribution mapping. A number of post-2003 records of more common millipedes were also added to the data set where they significantly extended the known distribution of the species. The habitat and other ecological data associated with these records were not included in the analysis.

Validation

The majority of the validation of the records has been carried out by the scheme organisers. As with the Centipede Recording Scheme, taxonomic validation has been based on the examination of initial voucher specimens from a new recorder followed by checking of further specimens considered problematical by the recorder. A small number of collectors have preferred to submit all specimens for identification by the scheme organiser. Validation of the geographical data on the RA13 and RA59 recording cards was also necessary. Ideally, on receipt of a card the grid reference, location name and vice-county were checked and cross referenced using OS maps. The most common mistakes (reversal of eastings/northings and misquoted 100 km square figures) were easily corrected but a small number of cards, where site names were not traceable from the maps, were either referred back to the card compiler or discarded.

Analysis

From its inception, an important objective of the Millipede Recording Scheme was to gather ecological data alongside the distributional data collected by all biological recording schemes. It was understood from the start that computers would be needed to process this data in order to enable information on habitat preferences and related ecological factors to be extracted and communicated to recorders in an understandable format. What could not be anticipated was the introduction, fifteen years later, of a new recording card that included changes to the habitat recording system. The existence of two versions of the habitat recording system has presented a number of problems in analysis of the data. These same problems were faced by the Centipede Recording Scheme but Barber and Keay (1988) were able to restrict their analysis of centipede records to the data from the original system on the RA14 cards (equivalent to the RA13 millipede recording cards) as relatively little data was contained on the new cards. However, the large amount of millipede data collected since the introduction of the RA59 card in 1985 could not be omitted from any meaningful analysis. Therefore an attempt has been made to combine the two datasets through identification of analogous fields within the two habitat recording systems. Wherever possible the habitat categories on the RA13 card (Figure 2) were mapped across to comply with the principal habitat categories on the RA59 card (Figure 3). The results of this exercise are given in Appendix 1.

However, strictly analogous categories were not always found and most of the more precise habitat divisions had to be combined in order to produce categories with sufficiently large numbers of records for valid analysis. Records where the principal habitat was given as "Other" were not used as it would have been necessary to check each of the 536 records to extract any meaningful information from them. Records on RA59 cards where the principal habitat was given as "Various" were also problematic. As these records were from a group of unrelated habitats it would have been impossible to produce a meaningful interpretation of any association between a species and the principal habitat. However, rather than discard all of the 1 512 records contained within this category, records for the three most important sub-categories were used. The sub categories "park/recreation" and "urban/suburban open space" were combined as a principal habitat and the sub-category "churchyards" was treated as a separate principal habitat. The principal habitat categories that were eventually used in the current analysis are shown in Table 3. The number of records for each species from each of these habitats is shown in Appendix 2. In addition to the principal habitat categories, an attempt was made to marry the other environmental and ecological data provided on the two recording cards. This proved to be even more problematic than the principal habitat data but it was considered worth analysing four other aspects of the ecological data for the full dataset.

Table 3: Number of records from each of the principal habitat categories used in the analysis of the millipede recording scheme data.

Principal habitat	Number of records
Building	647
Cave/Tunnel/Well	42
Cultivation	2 221
Excavation	843
Grassland	3 908
Heath/Moor	653
Maritime	5 254
Rock	618
Sand dune	538
Scrubland	1 403
Waste ground	2 791
Wetland	547
Woodland	16 317
Open space/Park/Recreation	969
Churchyard	482
TOTAL	**37 233**

These factors were soil/litter acidity (calcareous or non-calcareous), soil texture (heavy clay/clay or loam or peat or sand/pure sand), coastal/inland bias (less than 15 km from sea or more than 15 km from sea) and synanthropy (rural or suburban/urban bias). The number of records for each species from each of these categories is shown in Appendix 3.

A simple analysis of the dataset using the crude percentage of records of a species in a particular habitat is only valid if it can be assumed that each of the habitats are equally searched and each of the species are equally abundant. However, a problem common to the majority of recording schemes is that of varying recorder effort in different habitat types. A range of factors contribute to variations in recorder effort including the rarity of the habitat type, the ease of collecting in different habitats and the personal preferences of the collector. In addition, as with other organisms, the diversity of millipedes is not the same in the different habitats. Fairhurst (1984) recognised this problem in his analysis of the first 10 000 millipede records submitted on RA13 cards. However, he still chose to base his analysis on the assumption that each of the habitats are equally searched and each of the species are equally abundant due to the ease of interpretation of the simple percentages produced. Barber and Keay (1988) attempted a more sophisticated analysis of the centipede data through the use of a standardization procedure to address the problems outlined above.

The association between millipede species and habitat categories was analysed by comparing the number of observations of a species from a habitat category with its predicted number in that category. The relative scarcity of a species (as measured by the number of records for that species as a proportion of all records) was used as a weighting factor for the total number of records from a habitat category to predict the number of records of the species in each habitat category. Thus, taking the total number of records actually observed in each habitat category, the first step of the analysis was to ask the question, "how many records would we expect for a species in that habitat based on the relative scarcity of the species?"

The theoretical or expected value for the number of records of a species from a habitat category was calculated as:

$$\frac{\text{(total number of records of a species) x (total number of records from a habitat category)}}{\text{(total number of records of all species)}}$$

Using this expected value, the formula used to obtain the measure of association was:

$$\frac{\text{abs (observed value – expected value) x (observed value – expected value)}}{\text{expected value}}$$

where abs is the absolute value of the difference between observed and expected. This measure is effectively the contribution to a chi-square analysis of association, and gives the lack of fit to the marginal totals for species and habitats. It assesses the goodness of fit of the observed data to a theoretical model of the distribution of species records between the habitat categories. The absolute value of the difference between observed value and expected value is always a positive number. This allows positive associations to be distinguished from negative associations, something that is not possible with a normal Chi-squared analysis or with the standardization procedure adopted by Barber & Keay (1988). The results of the analyses are given in Table 4 and Table 5.

The measure of association calculated has the advantage of using information on both the frequency of sampling different habitats and differing species abundances. The major drawback is that there is no objective way of deciding what constitutes a significant positive or negative association. A subjective assessment had to be made of the level of significance of the calculated measure of association. This judgement had to be based on a comparison of anecdotal evidence of the habitat requirements of some of the better known species with the figures obtained by the chi-squared calculations. On this basis a threshold value of +/-20 was identified as indicating a strong relationship and +/-100 indicating a very strong relationship. However, where the total number of records for a species is low, associations with more frequently sampled habitats may not be made apparent whilst dubious associations with rarely sampled habitats may appear very strong. For example, at least 28 records of a species are required to indicate a significant association with woodland whereas a single record can indicate a significant association with buildings. For this reason, any interpretation of the measure of association for those species with less than 30 records needs to be treated with caution.

Table 4: Measures of association between each species of millipede and the principal habitat categories.
Values of at least +/-20 indicate a strong relationship. These are shown in bold type.
Values of at least +/-100 indicate a very strong relationship. These are shown in bold type against a grey highlight.

Species	Building	Cave	Cultivation	Excavation	Grassland	Heath/Moor	Maritime	Rock	Sand dune	Scrubland	Waste ground	Wetland	Woodland	Churchyard	Open space
Adenomeris gibbosa	0	0	7	0	0	0	2	0	0	0	0	0	-1	0	0
Allajulus nitidus	1	0	**21**	0	2	-2	-1	0	-1	0	-2	-1	-8	**28**	**26**
Anamastigona pulchella	0	0	0	0	-1	0	-1	0	0	0	0	0	4	0	0
Anthogona britannica	0	0	-1	0	0	0	5	2	0	-1	-1	0	0	0	0
Archiboreoiulus pallidus	2	0	**37**	-1	0	-4	-15	-1	-3	-2	-1	-3	1	11	0
Blaniulus guttulatus	2	2	**215**	0	0	-9	-4	0	-8	0	**21**	-13	**-80**	**174**	**36**
Boreoiulus tenuis	0	0	**26**	-2	-4	-2	-2	0	-3	9	0	-3	-14	**224**	18
Brachychaeteuma bagnalli	0	0	**31**	4	1	-1	0	2	-1	0	-1	-1	-7	7	-1
Brachychaeteuma bradeae	-1	0	**23**	-1	0	-1	0	-1	-1	-1	1	-1	-1	5	0
Brachychaeteuma melanops	0	0	**29**	0	0	-2	-4	0	-1	2	15	0	-10	5	7
Brachydesmus superus	-1	8	1	0	18	-8	-8	-2	-16	2	**49**	4	-5	-4	-1
Brachyiulus pusillus	0	-1	**21**	2	**33**	-6	7	0	13	0	8	8	**-80**	3	4
Choneiulus palmatus	**135**	0	1	0	-2	0	4	0	-1	0	13	0	**-28**	9	3
Chordeuma proximum	-5	0	-5	1	-13	0	-1	-1	-1	0	-5	0	**32**	-4	0
Chordeuma sylvestre	0	0	-1	0	-1	0	6	0	0	0	1	0	0	0	0
Craspedosoma rawlinsii	-1	0	-3	0	-5	0	0	-1	-1	-1	-3	**50**	7	-2	-1
Cylindroiulus britannicus	5	0	4	0	0	-9	-5	-1	-7	8	0	5	-1	2	12
Cylindroiulus caeruleocinctus	2	-1	**228**	0	13	-1	-18	-8	-7	1	16	-3	**-90**	**164**	**20**

Species	Building	Cave	Cultivation	Excavation	Grassland	Heath/Moor	Maritime	Rock	Sand dune	Scrubland	Waste ground	Wetland	Woodland	Churchyard	Open space
Cylindroiulus latestriatus	0	0	**-42**	-7	**28**	12	**595**	3	**1968**	-14	**-37**	-2	**-293**	**-21**	-4
Cylindroiulus londinensis	-1	0	10	0	0	-1	-1	-1	2	-1	0	-1	0	-1	-1
Cylindroiulus parisiorum	0	0	7	2	3	-1	-3	-1	-1	0	15	-1	-5	0	-1
Cylindroiulus punctatus	-13	-2	**-59**	-9	**-27**	1	-10	0	**-38**	0	**-29**	-1	**186**	**-25**	-7
Cylindroiulus salicivorus	**445**	0	0	0	-1	0	-1	0	0	0	-1	0	-4	0	0
Cylindroiulus truncorum	**59**	0	5	0	-1	0	-1	0	0	0	0	0	-4	15	0
Cylindroiulus vulnerarius	**32**	0	**57**	-1	-4	-1	-1	-1	-1	0	2	-1	-10	0	**28**
Enantiulus armatus	-1	0	2	-1	-3	-1	4	-1	-1	1	-1	0	0	-1	-1
Geoglomeris subterranea	1	0	-4	**39**	0	0	-3	8	-1	0	-3	-1	1	-2	-1
Glomeris marginata	-16	-3	**-70**	10	-1	-8	2	**88**	**-25**	9	**-21**	-13	**46**	**-52**	**-20**
Haplopodoiulus spathifer	0	0	**46**	0	-1	0	-1	0	0	0	0	0	-2	0	14
Julus scandinavius	-1	-2	**-20**	-3	-3	12	0	0	18	-5	**-31**	2	**30**	-18	0
Leptoiulus belgicus	1	0	1	0	-4	-2	4	0	0	3	2	1	-1	-3	0
Leptoiulus kervillei	-1	0	-3	0	-9	-1	15	0	-1	1	-3	-1	3	-2	0
Macrosternodesmus palicola	0	0	**119**	-2	-4	-5	**-21**	-2	-6	0	7	-6	**-35**	**465**	**28**
Melogona gallica	-2	0	0	2	-11	-1	0	-1	0	-3	-3	0	12	1	3
Melogona scutellaris	0	0	2	2	-3	-4	-5	-1	-5	0	-5	-2	11	0	0
Melogona voigti	0	0	0	0	0	0	0	0	0	0	0	0	1	0	0
Metaiulus pratensis	0	0	0	0	0	0	0	0	0	0	5	**32**	-1	0	0
Nanogona polydesmoides	0	**40**	1	17	-2	-11	1	0	-13	0	**45**	-10	-7	2	0
Nemasoma varicorne	-7	-1	-11	-3	0	-7	-7	-3	-6	0	**-25**	0	**55**	-4	1

Species	Building	Cave	Cultivation	Excavation	Grassland	Heath/Moor	Maritime	Rock	Sand dune	Scrubland	Waste ground	Wetland	Woodland	Churchyard	Open space
Nopoiulus kochii	**46**	0	-2	3	-1	0	-2	0	0	-1	**63**	0	-8	-1	8
Ommatoiulus sabulosus	0	0	**-23**	-2	-2	**54**	**69**	2	**240**	-6	-18	**64**	-14	**-26**	-5
Ophiodesmus albonanus	1	0	**171**	0	-2	-5	-6	-3	-4	0	2	-4	**-37**	**140**	**46**
Ophyiulus pilosus	3	-1	10	2	-1	-2	-4	-3	0	0	0	8	0	-2	7
Oxidus gracilis	**1565**	0	-1	-1	-4	-1	-2	-1	0	-1	-3	0	-15	-1	0
Polydesmus angustus	0	-1	-11	1	19	**80**	-2	-4	-3	-1	1	2	0	**-23**	-3
Polydesmus barberii	0	0	0	0	-1	0	-1	0	0	0	-1	0	0	15	**35**
Polydesmus coriaceus	-4	0	**96**	0	**26**	-8	-11	-8	-6	4	**82**	0	**-48**	1	1
Polydesmus denticulatus	-1	0	-16	-3	0	1	0	0	-2	-1	-5	**58**	9	-6	-2
Polydesmus inconstans	1	1	2	1	**23**	1	10	0	**22**	-1	1	1	**-39**	0	-3
Propolydesmus testaceus	0	0	0	1	1	0	12	0	0	0	0	0	-4	0	0
Polyxenus lagurus	**33**	0	0	-2	1	0	**64**	0	-2	-6	-2	-2	**-52**	**167**	-1
Polyzonium germanicum	0	0	0	0	0	0	-1	0	0	0	-1	0	3	0	0
Poratia digitata	**167**	0	0	0	0	0	0	0	0	0	0	0	-1	0	0
Prosopodesmus panporus	**56**	0	0	0	0	0	0	0	0	0	0	0	0	0	0
Proteroiulus fuscus	-13	0	**-42**	-13	**-39**	1	-4	-2	**-27**	0	**-56**	1	**180**	**-21**	-1
Stosatea italica	0	0	4	0	-4	-1	5	0	-1	-1	0	-1	0	-1	1
Tachypodoiulus niger	0	0	-5	5	13	0	-14	4	-10	0	**44**	-3	-1	-2	-8
Thalassisobates littoralis	0	0	-1	0	-2	0	**103**	1	0	-1	-2	0	-10	-1	0
Trachysphaera lobata	0	0	0	0	0	0	-1	0	0	5	0	0	1	0	0
Unciger foetidus	0	0	**20**	0	-1	0	5	0	0	0	0	0	-3	0	0

Table 5: Measures of association between each species of millipede and selected environmental factors.
Values of at least +/-20 indicate a strong relationship. These are shown in bold type.
Values of at least +/-100 indicate a very strong relationship. These are shown in bold type against a grey highlight.

Species	Non-Calcareous	Calcareous	Loam	Peat	Sand	Clay	Coastal	Inland	Rural	Urban / Suburban
Adenomeris gibbosa	0	0	0	0	0	0	1	-1	-1	3
Allajulus nitidus	-5	4	1	-2	2	-2	0	0	-7	**28**
Anamastigona pulchella	1	-1	0	0	0	0	1	-1	0	0
Anthogona britannica	2	-2	0	0	0	0	**22**	-10	1	-2
Archiboreoiulus pallidus	-1	1	6	-7	-9	0	-10	5	-6	**24**
Blaniulus guttulatus	-17	14	13	-18	-12	0	0	0	**-72**	**293**
Boreoiulus tenuis	0	0	0	-4	-7	14	0	0	**-40**	**161**
Brachychaeteuma bagnalli	-3	3	-3	-2	-2	16	4	-2	-8	**33**
Brachychaeteuma bradeae	0	0	-1	0	0	1	-1	0	-4	17
Brachychaeteuma melanops	-5	4	1	-3	-5	1	8	-4	-16	**66**
Brachydesmus superus	-12	10	-3	0	-12	**23**	0	0	2	-7
Brachyiulus pusillus	-8	6	**-24**	0	5	**24**	**23**	-11	-2	7
Choneiulus palmatus	0	0	0	0	3	-1	11	-5	-13	**51**
Chordeuma proximum	9	-7	1	-1	-1	0	**30**	-14	0	-2
Chordeuma sylvestre	1	-1	1	0	0	-1	7	-3	0	-1
Craspedosoma rawlinsii	6	-5	-4	**50**	0	-1	0	0	1	-6
Cylindroiulus britannicus	-3	2	1	-3	-10	3	1	-1	-10	**42**
Cylindroiulus caeruleocinctus	**-35**	**29**	8	-14	1	-6	**-32**	15	**-57**	**233**

Species	Non-Calcareous	Calcareous	Loam	Peat	Sand	Clay	Coastal	Inland	Rural	Urban / Suburban
Cylindroiulus latestriatus	0	0	**-87**	-6	**930**	**-78**	**766**	**-351**	6	**-24**
Cylindroiulus londinensis	1	-1	-1	-1	11	0	-1	0	0	0
Cylindroiulus parisiorum	-2	1	0	0	0	0	-11	5	-1	4
Cylindroiulus punctatus	**28**	**-23**	5	3	-9	-2	-19	9	**21**	**-87**
Cylindroiulus salicivorus	0	0	0	0	0	0	12	-5	-6	**26**
Cylindroiulus truncorum	1	-1	-1	**25**	0	-1	0	0	-2	9
Cylindroiulus vulnerarius	-2	1	0	-1	0	0	0	0	**-25**	**101**
Enantiulus armatus	2	-1	1	0	0	-3	14	-7	0	1
Geoglomeris subterranea	-16	13	1	0	-3	0	-4	2	2	-7
Glomeris marginata	**-46**	**37**	**26**	-10	-10	-7	-10	5	**37**	**-151**
Haplopodoiulus spathifer	0	0	1	0	0	0	-2	-2	-1	3
Julus scandinavius	19	-15	1	0	12	-12	6	-3	1	-4
Leptoiulus belgicus	7	-5	2	-3	2	-3	**91**	**-42**	0	0
Leptoiulus kervillei	15	-12	17	-3	-4	-8	16	-7	1	-3
Macrosternodesmus palicola	-19	16	-1	-9	-2	14	-4	2	**-49**	**201**
Melogona gallica	3	-3	9	-1	-2	-5	6	-3	-2	7
Melogona scutellaris	0	0	0	-2	-8	7	-7	3	-7	**27**
Melogona voigti	0	0	0	0	0	0	0	0	0	0
Metaiulus pratensis	0	0	1	0	0	0	0	0	-2	7
Nanogona polydesmoides	-5	4	0	-7	-2	4	0	0	-6	**24**
Nemasoma varicorne	-1	1	-3	0	-3	10	**-21**	10	7	**-28**

Species	Non-Calcareous	Calcareous	Loam	Peat	Sand	Clay	Coastal	Inland	Rural	Urban / Suburban
Nopoiulus kochii	0	0	-1	-1	16	-2	3	-2	-13	**52**
Ommatoiulus sabulosus	16	-13	-15	**62**	**172**	**-69**	**60**	**-28**	18	**-75**
Ophiodesmus albonanus	-15	12	3	-7	-14	3	0	0	**-45**	**184**
Ophyiulus pilosus	0	0	1	0	-17	3	3	-2	-1	4
Oxidus gracilis	1	-1	0	**27**	-1	-1	1	0	-17	**71**
Polydesmus angustus	**29**	**-23**	-2	**35**	1	-3	-14	7	2	-9
Polydesmus barberii	0	0	0	0	0	0	12	-5	0	2
Polydesmus coriaceus	**-21**	17	-16	-5	**-29**	**102**	**-27**	12	-4	16
Polydesmus denticulatus	1	-1	-4	0	0	7	0	0	6	**-23**
Polydesmus inconstans	-2	1	-1	0	4	0	6	-3	0	-1
Propolydesmus testaceus	-3	2	2	0	0	-1	6	-3	0	0
Polyxenus lagurus	-2	1	-1	-1	0	2	12	-5	-2	7
Polyzonium germanicum	0	0	3	0	-1	-2	-3	1	0	-2
Poratia digitata	1	-1	0	13	0	0	-1	0	-2	10
Prosopodesmus panporus	0	0	0	13	0	0	0	0	-1	3
Proteroiulus fuscus	**60**	**-49**	-9	**21**	2	1	-1	1	10	**-41**
Stosatea italica	-4	3	2	-1	-2	0	2	-1	-1	3
Tachypodoiulus niger	0	0	8	0	-15	-1	**-69**	**32**	4	-18
Thalassisobates littoralis	0	0	-2	0	19	-1	**34**	-16	0	-2
Trachysphaera lobata	3	-2	-2	0	19	-1	6	-3	0	-1
Unciger foetidus	1	-1	0	0	2	-1	4	-2	-2	10

Conservation

Bratton (1991) assigned RDB K, Insufficiently Known status to just two species of millipede, *Trachysphaera lobata* and *Chordeuma sylvestre*. A review of the conservation status of the British millipedes has never been published but nationally notable B (= nationally scarce) status is accorded a number of species in the Recorder 3.3 species database and in the absence of anything else this has been used as a standard reference by those working in conservation. The species accorded nationally scarce status are:-

- *Brachychaeteuma bagnalli*
- *Brachychaeteuma bradeae*
- *Brachychaeteuma melanops*
- *Enantiulus armatus*
- *Nopoiulus kochii*
- *Polydesmus testaceus* (=*Propolydesmus testaceus*)
- *Polyzonium germanicum*
- *Stosatea italica*
- *Thalassisobates littoralis*

The UK Biodiversity Action Plan process initially resulted in the inclusion of seven species of millipede on the UK BAP Long List (Anon., 1995). The species included on this list were:-

- *Chordeuma proximum*
- *Chordeuma sylvestre*
- *Melogona scutellare* (= *M. scutellaris*)
- *Metaiulus pratensis*
- *Nanogona polydesmoides*
- *Polydesmus gallicus* (= *P. coriaceus*)
- *Trachysphaera lobata*

Harding (1998) explained the rationale behind the selection of these species. The criteria used in the selection process included consideration of the global importance of UK populations in addition to their scarcity in the UK. Insufficient data were available to apply criteria based on declines and threats. The application of the 'global importance' criterion resulted in a list very different from that based on scarcity alone. Most notable was the inclusion of *Nanogona polydesmoides*, a very common animal in Britain but of very restricted occurrence in continental Europe. When the UK BAP was reviewed in 1998 none of the millipedes were upgraded to 'Priority Species' status (Anon., 1998). All seven species are currently included in the 'Species of conservation concern' (SoCC) category. A review of the Priority species included in the UK BAP, using IUCN criteria (IUCN, 2001), was begun in 2004. Six millipedes were proposed for Priority species status in the initial stages of the review and three have been accepted for further consideration. The three candidates for Priority species status are:-

- *Metaiulus pratensis*
- *Polyzonium germanicum*
- *Trachysphaera lobata*

Metaiulus pratensis and *P. germanicum* were both proposed on the basis of meeting criteria for rarity and decline in Britain. The British population of *M. pratensis* is also of international significance. *Trachysphaera lobata* is seriously threatened at its only British site and along with *M. pratensis* seems to meet IUCN criteria for a critically endangered species based on restricted occurrence and declines in population size / geographic range. A JNCC led review of the conservation status of the whole of our millipede fauna is now overdue. The IUCN criteria need to be applied to each species to reassess RDB and nationally scarce statuses. Use of the data collected by the millipede recording scheme will be essential in undertaking this review.

Future recording

The production of a distribution atlas can be seen as the end product of a recording scheme. Years of voluntary activity have culminated in a publication which allows an individual recorder to identify their contribution, in the form of dots on a map, to the group effort. Personal gratification of this kind is an important incentive when encouraging field naturalists to spend their valuable leisure time filling in record cards rather than undertaking further field work. The many recorders who have contributed to this volume can be justifiably proud of their achievements. However, there is still much to be discovered and few recorders would have it any other way.

A distribution map can only ever provide a 'still picture' of a fluid situation. The ranges of some species are changing more rapidly than others, be they expanding as appears to be the case with *Chordeuma proximum*, or declining as with *Polyzonium germanicum*. In either case the distribution maps in this volume provide a baseline against which to monitor future changes. However, they can only be used in this way if naturalists in general and BMIG members in particular continue to record the distribution of millipedes as least as intensively as they have in the past. Therefore it is important that BMIG and the recording scheme not only continue to recruit new members but also maintain and improve the support provided to recorders.

Rather than undergoing rapid change, the ranges of most species may only be fluctuating at the edges as their most exposed populations are influenced by the interplay of a range of natural and man-made factors. The habitat analyses in this volume have provided a broad idea of what these important factors limiting distribution may be for part of our millipede fauna. However, there is still much work to be done. There are still species, such as *Adenomeris gibbosa*, where almost nothing is known of their habitat preferences. For most species, even though we now have some indication of their ecological preferences, we still need much more detailed information on the key factors governing their distributions. Detailed autecological studies may be necessary to obtain such information for the rare species but more detailed information for many of our species can be gathered by millipede recorders consistently providing as much habitat data as possible when submitting their records. In the course of analysing the data for this volume it became apparent that there are problems with the existing RA59 card which will need to be addressed before the next phase of millipede recording in Britain and Ireland is launched. Any novice recorders who have struggled with the complexities of the RA59 card can take heart from the example of the very experienced recorder who classified exactly the same site as two totally different habitats on different dates.

Following publication of this atlas the millipede dataset will no longer be managed by BRC. As is the case with a number of other recording schemes, e.g. the Spider Recording Scheme, BMIG will take over responsibility for the existing dataset and for the collection, input and analysis of all future millipede records. The details of this have yet to be finalised but it will probably involve encouraging recorders to submit their records electronically and the use of a commercially available software package to manage the recording scheme data. This is one area where support for recorders is going to be essential but other recording schemes have already proved how successful the approach can be. One advantage of electronic records that should be appreciated by individual recorders will be the opportunity to update distribution maps much more frequently than in the past. Negotiations are underway with officers of the National Biodiversity Network to make the existing dataset available through the NBN Gateway and to agree a schedule for the update of the distribution maps and underlying data in the future.

In addition to the site based recording that has been the mainstay of the millipede recording scheme for the past 30 years, there is a need for more targeted recording in future. It is apparent from Appendix 2 that some habitats are seriously under-worked. Caves are the most obvious example of this. A review of the literature shows that caves are an important habitat for millipedes in continental Europe. Some of our species are found more frequently in caves than above ground on the continent and some species have only been found in caves. Despite this there have been only 42 British and Irish records from this habitat over the lifetime of the recording scheme. Clearly there is a need for targeted recording of caves in the next phase of the recording scheme. This will require input from recorders with specialist skills but there are other under-worked habitats that can be targeted by all recorders. These include habitats commonly visited by naturalists on field meetings such as heathland, moorland, sand dunes and wetlands.

Targeted surveys of specific species are also required. Some recently discovered species such as *Trachysphaera lobata* and *Adenomeris gibbosa* appear to have very restricted distributions. There are no obvious reasons for this and it may be that we simply do not know where and how to look for these millipedes as yet. Further survey work is required firstly to determine whether or not these species are truly as limited in distribution as they appear to be and secondly to determine what their habitat requirements actually are. In some cases species are under-recorded because specialist techniques are required to find them. This description applies to relatively common species such as *Polyxenus lagurus* as well as to rarities such as *Thalassisobates littoralis*. Another example of a need to undertake surveys of individual species is demonstrated by *Melogona gallica* s.l. Now that *Melogona voigti* has been recognised as a British species there is an urgent need to revisit the old sites for *Melogona gallica* s.l. in order to determine the true distribution of the two species.

Finally it was obvious from the data analysed that most millipede recorders are active in the spring and early summer but very few bother to record throughout the year. This has created a bias in the recording scheme data as many species are adult and active through the winter months. There is a need for more targeted recording at this time of year in order to obtain a clearer picture of the relative scarcity of different species and to produce a more complete species list for any site being recorded.

The unfamiliarity of millipedes to most naturalists, never mind to the general public, was referred to in the introduction to this volume. In spite of this handicap a relatively small number of committed recorders have already achieved a great deal in amassing the wealth of data upon which the maps and analyses are based. The atlas may have provided few definitive answers in itself but it has enabled many questions to be posed more clearly. It has also provided a very strong foundation of baseline data upon which the next phase of the recording scheme can build. With the enthusiasm that millipede recorders have shown so far there is no reason to suppose that the answers to many of these questions will not be discovered as the second phase of recording gets underway.

Appendix 1

Combination of habitat categories from RA13 and RA59 recording cards

The existence of two versions of the habitat recording system for millipedes presented a number of problems in analysis of the recording scheme data. An attempt was made to combine the two datasets through identification of analogous fields within the two habitat recording systems. Wherever possible the habitat categories on the RA13 card (Figure 2) were mapped across to comply with the principal habitat categories on the RA59 card (Figure 3). The results of this exercise are given on the following pages.

LOCALITY (22–24)

MILLIPEDES 6100–6119 (1–4)

Date (60–64) | V.C. No. (33–35)

(77–79) V.C.

RECORDER

DET.

Alt. (73–76) | Code No (65–68)

Grid Ref. (25–32)

(5–10)
101 Archeboreoiulus pallidus
201 Blaniulus guttulatus
301 Boreoiulus tenuis
401 Brachychaeteuma bagnalli
402 bradeae
403 melanops
501 Brachydesmus superus
601 Brachyiulus pusillus
701 Choneiulus palmatus
801 Chordeuma proximum
802 silvestre
901 Craspedosoma rawlinsi
1001 Cylindroiulus britannicus
1002 latestriatus
1003/1 londinensis var typica
1003/2 l. var caeruleocinctus
1003/3 l. var finitimus
1004 nitidus
1005 parisiorum
1006 punctatus
1101 Entothalassinium italicum
1201 Eumastigonodesmus bonci
1301 Geoglomeris jurassica
1401 Glomeris marginata
1501 Isobates littoralis
1502 varicornis

(5–10)
1601 Iulus scandinavius
1701 Leptoiulus belgicus
1702 kervillei
1801 Leptophyllum armatum
1901 Macrosternodesmus palicola
2001 Metaiulus pratensis
2101 Microchordeuma gallicum
2102 scutellare
2201 Nopoiulus minutus
2301 Ommatoiulus (= Schizophyllum) sabulosum
2401 Ophoidesmus albonanus
2501 Ophyiulus pilosus
2601 Oxidus gracilis
2701 Polydesmus angustus
2702 coriaceus
2703 denticulatus
2704 gallicus
2705 testaceus
2801 Polymicrodon polydesmoides
2901 Polyxenus lagurus
3001 Polyzonium germanicum
3101 Proteroiulus fuscus
3201 Tachypodoiulus niger

Other species:

HABITAT DATA

A 1 tick (obligatory):	11
Coastal <15km from sea	1
Inland >15km from sea	2

B 1 tick (obligatory):	12
Urban	1
Suburban/village	2
Rural	3

C 1st order habitats; 1 tick (obligatory):	13–15
Aquatic: Canal	001
River >5m wide	002
Lake >1 acre (0.4 hectare)	003
Estuary	004
Sea	005
Marsh: Fen	011
Carr	012
Bog	013
Salt marsh	014
Cave/Well/Tunnel: Threshold	021
Dark zone	022
Building: Inside	031
Outside	032
Garden: Domestic	041
Waste ground: <25% veg. cover	051
>25% veg. cover	052
Arable: Cereal crops	061
Root crops	062
Fodder crops	063
Grass ley	064
Market garden/allotment	065
Grassland: Ungrazed	071
Lightly grazed	072
Heavily grazed	073
Mown	074
Scrubland: Dense	081
Open with herbs/grass	082
Woodland: Dense	091
Open with scrub	092
Open with herbs/grass	093
Acid heath/moor: Moss/lichen	101
Grass/sedge/rush	102
Heather	103
Vaccinium (bilberry)	104
Mixed	105
Sand dune: Bare sand	201
Tussocky	202
Dense sward	203
Dune slack	204
Dune heath	205
Other: If none of above	301

D 2nd order habitats; 1 tick (where applicable):	16, 17
Cold frame	01
Rockery	02
Flower bed	03
Lawn	04
Compost/refuse heap	05
Dung heap	11
Hay (or other) stack	12
Potato (or other) clamp	13
Hedge	21
Roadside verge	22
Embankment/cutting	23
Woodland ride/firebreak	24
Wood fence	25
Dry stone wall	31
Wall with mortar	32
Quarry face	33
Quarry floor	34
Natural cliff face	35
Rock pavement	36
Stablilised scree	37
Unstabilised scree	38
Grike	41
Road/path	51
Dry water course bed	61
'Dry' ditch	71
Wet ditch	72
Shore/water edge/strandline	81
Vegetated stream	91
Unvegetated stream	92
Puddle	93
Pond <1 acre (0.4 hectare)	94
Flood patch	95

Biological Records Centre June 1970 RA 13

Figure 2: The original (RA13) millipede recording card.

HABITAT DATA

E MICROSITE (animal actually found under, on or in); 1 tick (obligatory): 36, 37

Stones	01
Shingle	02
Soil/sand	03
Litter	11
Tussocks	21
Bark (living trees or shrubs)	31
Dead wood	32
Dung	33
Carrion	34
Bracket fungi	35
Ant colony (specify if possible)	41
Bird/mammal nest (specify below)	51
Rock	61
Stone or brick work	62
Shore line jetsam	71
Human rubbish/garbage	81
Other (specify below if possible)	91

F HABITAT QUALIFIERS; 1 tick in each section, where applicable:

	38
(a) Building: Cellar	1
Inhabited/public	2
Uninhabited/outbuilding	3
Ruin	4
Greenhouse (heated)	5
Greenhouse (unheated)	6

	39
(b) Shore: Intertidal	1
Splash zone	2
Between splash zone and 100m	3
100–1000m above H.W.M.	4

	40
(c) Encrustations: Moss	1
Lichen	2
Pleurococcoids	3

	41
(d) Waterspeed: Fast	1
Slow	2
Standing	3

	42
(e) Watercourse bed: Rocks	1
Pebbles	2
Sand	3
Silt	4
Peat	5

G LIGHT LEVEL; 1 tick (obligatory):

	43
Full daylight	1
Half-light/dusk/dawn	2
Dark	3

Other information e.g.
Abundance,..................(number) collected per..................(please state unit of time/space/volume) (53) Aspect/degree of slope (54) Behaviour (55) Food (56) Predators and Parasites (57) Age structure and Sex ratio (58) etc. (59)

H SOIL/LITTER DETAILS, terrestrial habitats only; 1 tick in each section where applicable:

	44, 45
(a) Litter mainly: Oak	01
Beech	02
Birch	03
Sycamore	04
Mixed deciduous	05
Coniferous	11
Mixed decid./conif.	21
Gorse	31
Hawthorn	32
Heathers	33
Sea Buckthorn	34
Litter/veg. mainly: *Carex*	41
Molinia	42
Dactylis	43
Festuca	44
Bromus	45
Brachypodium	46
Grass—species unknown	47
Mixed grass/herbs	51
Nettles	61
Reeds (*Phragmites*)	62
Juncus	63
Bracken	64
Other (specify below)	71

	46
(b) Litter age: Fresh	1
Old	2
Both	3

	47
(c) Litter cover: Exposed	1
Protected by thin veg.	2
Protected by thick veg.	3

	48
(d) Soil/exposed rock: Calcareous	1
Non-calcareous	2

	49
(e) Soil: Heavy clay	1
Clayey	2
Peat	3
Loam	4
Sandy	5
Pure sand	6

	50
(f) Humus type: *Mull*	1
Mor	2

I LOCATION OF ANIMAL; 1 tick in each section where applicable:

	51
(a) Horizon: >3m above ground	1
<3m above ground	2
On ground surface	3
In litter	4
<10cm in soil	5
>10cm in soil	6

	52
(b) Position: In open	1
In crevice	2

(8-15) Grid ref.

Vice-county (5-7)

Locality

MILLIPEDES 6100 (1-4)

Recorder/Collector	Determiner	Compiler	Date (30-35)
(18-21)	(22-25)	(26-29)	Altitude (36-39) metres
No.	No.	No.	Source (40): Field / Mus. / Lit.

? (41-42)	♂ (43-44)	♀ (45-46)	ϴ (47-48)	(49-53)	
				3301	Adenomeris gibbosa
				101	Archiboreoiulus pallidus
				201	Blaniulus guttulatus
				301	Boreoiulus tenuis
				401	Brachychaeteuma bagnalli
				402	bradeae
				403	melanops
				501	Brachydesmus superus
				601	Brachyiulus pusillus
				701	Choneiulus palmatus
				801	Chordeuma proximum
				802	sylvestre
				901	Craspedosoma rawlinsii
				1001	Cylindroiulus britannicus
				1007	caeruleocinctus
				1002	latestriatus
				1003	londinensis
				1004	nitidus
				1005	parisiorum
				1006	punctatus
				1008	truncorum
				1009	vulnerarius
				1801	Enantiulus armatus
				1201	Eumastigonodesmus bonci
				1401	Glomeris marginata
				1601	Julus scandinavius
				1701	Leptoiulus belgicus
				1702	kervillei
				1901	Macrosternodesmus palicola
				2101	Melogona gallica
				2102	scutellare
				2001	Metaiulus pratensis
				2801	Nanogona polydesmoides
				1502	Nemasoma varicorne
				2201	Nopoiulus kochii
				2301	Ommatoiulus sabulosus
				2401	Ophiodesmus albonanus
				2501	Ophyiulus pilosus
				2601	Oxidus gracilis
				2701	Polydesmus angustus
				2703	denticulatus
				2704	gallicus
				2702	inconstans
				2705	testaceus
				2901	Polyxenus lagurus
				3001	Polyzonium germanicum
				3401	Prosopodesmus panporus
				3101	Proteroiulus fuscus
				1101	Stosatea italica
				1301	Stygioglomeris crinita
				3201	Tachypodoiulus niger
				1501	Thalassisobates littoralis
				3501	Unciger foetidus

Other species:

Biological Records Centre December 1984 RA59

Figure 3: The revised (RA59) millipede recording card.

PRINCIPAL HABITATS

Tick one only

(55-57)

010	MARITIME
011	rock
012	shingle
013	sand
014	mud/silt
015	saltmarsh
016	artificial construction
017	sea cliff
020	SAND DUNE
021	bare sand
022	tussocky
023	dense sward
024	dune slack
025	dune heath
026	machair
030	WETLAND
031	fen
032	carr
033	bog
040	HEATH/MOOR mainly
041	moss/lichen
042	grass/sedge/rush
043	heather
044	Vaccinium
045	bracken
046	juniper
047	gorse
048	mixed
049	other (specify)
050	GRASSLAND
051	ungrazed
052	lightly grazed
053	heavily grazed
054	mown
060	VARIOUS
061	park/recreation
062	orchard
063	tree nursery
064	young plantation
065	hop field
066	soft fruit field
067	vineyard
068	churchyard
069	urban/suburban open space
070	CULTIVATION
071	grass ley
072	cereal crops
073	other seed crops
074	root crops
075	Brassica/other veg
076	flower crops
077	garden, ornamental
078	garden, domestic
080	SCRUBLAND
081	dense
082	open with grass/herbs
	WOODLAND mainly
090	mixed deciduous
091	poplar/aspen
092	birch
093	alder
094	hornbeam
095	hazel
096	beech
097	chestnut
098	oak
099	elm
101	acacia
102	field maple
103	sycamore
104	lime
105	ash
110	mixed coniferous
111	fir
112	spruce
113	pine
114	larch
115	yew
120	mixed coniferous/deciduous
130	other (specify)
140	BUILDING
141	outside
142	ruin
143	inside, uninhabited
144	inside, inhabited
145	glasshouse, unheated
146	glasshouse, heated
150	ROCK
151	flat/pavement
152	scree
153	inland cliff
160	CAVE/TUNNEL/WELL
161	threshold
162	dark zone
170	EXCAVATION
171	sand pit
172	gravel pit
173	opencast mine
174	quarry floor
175	quarry face
176	scree
180	WASTE GROUND
181	demolition site
182	spoil tip (specify type)
	organic/refuse tip
190	OTHER (specify)

HABITAT DETAILS

Tick one in each section

(58)

1	Intertidal
2	Around High Water Mark
3	Splash zone
4	Coastal (< 1 km from sea)
5	Coastal (1-10 km from sea)
6	Inland (> 10 km from sea)

(59)

1	Urban
2	Suburban/Village
3	Rural

ECOTONE (60-61)

10	Waterside: pond
11	lake
12	ditch/stream
13	river
14	canal
15	saline
20	Road/railside verge
21	Path and verge
30	Woodland edge/ride
40	Tree row
41	Hedge
50	Wall, with mortar
51	Wall, drystone
52	Wood fence
60	Other (specify)

MICROSITE (62-64)

LOCATION (62)

01	Under
02	On
03	In

SUBSTRATE (63-64)

01	Rock
02	Stone
03	Shingle
04	Soil/sand
05	Leaf/twig litter
06	Tussocks/clumps
07	Bark surface
08	Dead wood
09	Dead wood/under bark
10	Moss/Lichen
11	Dung heaps
12	Carrion
13	Fungi
14	Nest (specify)
15	Shore line jetsam
16	Human rubbish/garbage
17	Other (specify)

SLOPE (65)

1	Flat
2	Shallow (5-15°)
3	Definite (15-25°/1 in 4)
4	Steep (25-45°/1 in 2)
5	Very Steep (> 45°/1 in 1)

ASPECT (66)

1	Flat
2	Facing N
3	Facing E
4	Facing S
5	Facing W

SOIL (67-68)

01	Heavy clay
02	Clayey
03	Peat
04	Loam
05	Sandy
06	Pure sand
11	Shallow (< 10 cm)

SOIL/ROCK (69)

1	Calcareous
2	Non-Calcareous

METHOD OF COLLECTION (70)

1	Casual/turnover
2	Dig/Sieve/Sort
3	Pitfall trap
4	Extraction
5	Existing collection

COMMENTS eg

(71)	Local topography (eg top/bottom of slope)
(72)	Soil moisture status
(73)	Litter type
(74)	Behaviour
(75)	Food
(76)	Predators/parasites
(77)	Etc
(80)	Site conservation status

RA59		RA13				
PRINCIPAL HABITATS		1st ORDER HABITATS C (also A & B)	2nd ORDER HABITATS D	MICROSITE E	HABITAT QUAL F	SOIL/LITTER H
010	MARITIME					
011	rock					
012	shingle					
013	sand					
014	mud/silt					
015	saltmarsh	014 saltmarsh				
016	artificial construction					
017	sea cliff	A1 coastal < etc.	35 natural cliff face			
020	SAND DUNE					
021	bare sand	201 bare sand				
022	tussocky	202 tussocky				
023	dense sward	203 dense sward				
024	dune slack	204 dune slack				
025	dune heath	205 dune heath				
026	machair					
030	WETLAND					
031	fen	011 fen				
032	carr	012 carr				
033	bog	013 bog				
040	HEATH/MOOR mainly					
041	moss/lichen	101 moss/lichen				
042	grass/sedge/rush	102 grass/sedge/rush				
043	heather	103 heather				
044	Vaccinium	104 Vaccinium (bilberry)				
045	bracken					
046	juniper					
047	gorse					
048	mixed	105 mixed				
049	other (specify)					

RA59		RA13				
PRINCIPAL HABITATS		1st ORDER HABITATS C (also A & B)	2nd ORDER HABITATS D	MICROSITE E	HABITAT QUAL F	SOIL/LITTER H
050	GRASSLAND					
051	ungrazed	071 ungrazed				
052	lightly grazed	072 lightly grazed				
053	heavily grazed	073 heavily grazed				
054	mown	074 mown				
060	VARIOUS					
061	park/recreation					
062	orchard					
063	tree nursery					
064	young plantation					
065	hop field					
066	soft fruit field					
067	vineyard					
068	churchyard					
069	urban/suburban open space					
070	CULTIVATION					
071	grass ley	064 grass ley				
072	cereal crops	061 cereal crops				
073	other seed crops					
074	root crops	062 root crops				
075	Brassica/other veg					
076	flower crops					
077	garden, ornamental					
078	garden, domestic	041 garden, domestic				
080	SCRUBLAND					
081	dense	081 dense				
082	open with grass/herbs	082 open with grass/herbs				

	RA59 PRINCIPAL HABITATS	RA13 1st ORDER HABITATS C (also A & B)	2nd ORDER HABITATS D	MICROSITE E	HABITAT QUAL F	SOIL/LITTER H
	WOODLAND mainly					
090	mixed deciduous	091/092/093 woodland				(a)05 mixed deciduous
091	poplar/ aspen					
092	birch	091/092/093 woodland				(a)03 birch
093	alder					
094	hornbeam					
095	hazel					
096	beech	091/092/093 woodland				(a)02 beech
097	chestnut					
098	oak	091/092/093 woodland				(a)01 oak
099	elm					
101	acacia					
102	field maple					
103	sycamore	091/092/093 woodland				(a)04 sycamore
104	lime					
105	ash					
110	mixed coniferous	091/092/093 woodland				(a)11 coniferous
111	fir					
112	spruce					
113	pine					
114	larch					
115	yew					
120	mixed coniferous/deciduous	091/092/093 woodland				(a)21 mixed decid/conif
130	other (specify)					
140	BUILDING					
141	outside	032 outside				
142	ruin				(a)4 ruin	
143	inside, uninhabited	031 inside			(a)3 uninhabited	
144	inside, inhabited	031 inside			(a)2 inhabited	
145	glasshouse, unheated				(a)6 greenhouse (unheated)	
146	glasshouse, heated				(a)5 greenhouse (heated)	

RA59		RA13				
PRINCIPAL HABITATS		1st ORDER HABITATS C (also A & B)	2nd ORDER HABITATS D	MICROSITE E	HABITAT QUAL F	SOIL/LITTER H
150	ROCK					
151	flat/pavement		36 rock pavement			
			41 grike			
152	scree		37 stabilised scree			
			38 unstabilised scree			
153	inland cliff	A2 inland etc.	35 natural cliff face			
160	CAVE/TUNNEL/WELL					
161	threshold	021 threshold				
162	dark zone		022 dark zone			
170	EXCAVATION					
171	sand pit					
172	gravel pit					
173	opencast mine					
174	quarry floor		34 quarry floor			
175	quarry face		33 quarry face			
176	scree					
180	WASTE GROUND					
		051 <25% veg cover				
		052 >25% veg cover				
181	demolition site					
182	spoil tip (specify type)					
	organic/refuse tip		05 compost / refuse heap			
			12 hay stack			
			13 potato clamp			
190	OTHER (specify)	301 Other				

RA59		RA13				
(58)		1st ORDER HABITATS C (also A & B)	2nd ORDER HABITATS D	MICROSITE E	HABITAT QUAL F	SOIL/LITTER H
1	Intertidal	A1 coastal < 15km from coast			(b)1 intertidal	
2	around HWM					
3	splash zone	A1 coastal < 15km from coast			(b)2 splash zone	
4	coastal (<1km from sea)	A1 coastal < 15km from coast			(b)3 btw. splash zoneand 100m	
					(b)4 100 to 1000m above HWM	
5	coastal (1-10km from sea)	A1 coastal < 15km from coast				
6	inland (>10km from sea)	A2 inland >15km from coast				

RA59		RA13				
(59)		1st ORDER HABITATS C (also A & B)	2nd ORDER HABITATS D	MICROSITE E	HABITAT QUAL F	.SOIL/LITTER H
1	urban	B1 urban				
2	suburban/village	B2 suburban/ village				
3	rural	B3 rural				

RA59			RA13				
ECOTONE			1st ORDER HABITATS C (also A & B)	2nd ORDER HABITATS D	MICROSITE E	HABITAT QUAL F	SOIL/LITTER H
10	Waterside:	pond		94 pond			
11		lake	003 lake > 1 acre				
12		ditch/stream		72 wet ditch			
				91 vegetated stream			
				92 unvegetated stream			
13		river	002 river > 5m wide				
14		canal	001 canal				
15		saline	004 estuary				
			005 sea				
20	road/railside verge			22 roadside verge			
				23 embankment / cutting			
21	path and verge						
30	woodland edge / ride			24 woodland ride / firebreak			
40	tree row						
41	hedge			21 hedge			
50	wall with mortar			32 wall with mortar	[62 stone or brick work]		
51	wall, drystone			31 drystone wall	[62 stone or brick work]		
52	wood fence			25 wood fence			
60	other (specify)						

RA59		RA13				
MICROSITE/SUBSTRATE		1st ORDER HABITATS C (also A & B)	2nd ORDER HABITATS D	MICROSITE E	HABITAT QUAL F	SOIL/LITTER H
01	rock			61 rock		
02	stone			01 stones		
03	shingle			02 shingle		
04	soil/sand			03 soil/sand		
05	leaf/twig litter			11 litter		
06	tussocks/clumps			21 tussocks		
07	bark surface			31 bark		
08	dead wood			32 dead wood		
09	dead wood/under bark					
10	moss/lichen				(c)1 moss	
					(c)2 lichen	
					(c)3 pleurococcus	
11	dung heaps		11 dung heap			
				33 dung		
12	carrion			34 carrion		
13	fungi			35 bracket fungi		
14	nest (specify)			41 ant colony		
				51 bird/mammal nest		
15	shore line jetsam			71 shore line jetsam		
16	human rubbish/garbage			81 human rubbish/garbage		
17	other (specify)			91 other (specify)		

RA59		RA13				
SOIL		1st ORDER HABITATS C (also A & B)	2nd ORDER HABITATS D	MICROSITE E	HABITAT QUAL F	SOIL/LITTER H
01	heavy clay					(e)1 heavy clay
02	clayey					(e)2 clayey
03	peat					(e)3 peat
04	loam					(e)4 loam
05	sandy					(e)5 sandy
06	pure sand					(e)6 pure sand
07	shallow (<10cm)					

RA59		RA13				
SOIL/ROCK		1st ORDER HABITATS C (also A & B)	2nd ORDER HABITATS D	MICROSITE E	HABITAT QUAL F	.SOIL/LITTER H
1	calcareous					(d)1 calcareous
2	non-calcareous					(d)2 non-calcareous

NO DIRECT EQUIVALENCE	RA13		
	1st ORDER HABITATS C	2nd ORDER HABITATS D	HABITAT QUAL F
	063 fodder crops 065 market garden /allotment	01 cold frame 02 rockery 03 flower bed 04 lawn 51 road/ path 61 dry water course bed 71 dry ditch 81 shore/ water edge/ starndline 93 puddle 95 flood patch	(a)1 cellar (d) waterspeed - all 3 categories (e) water course bed - all 5 categories
	LIGHT LEVEL G	.SOIL/LITTER H	.SOIL/LITTER (cont.) H
	all 3 categories	(a)31 gorse (a)32 hawthorn (a)33 heather (a)34 sea buckthorn (a)41 carex (a)42 Molina (a)43 Dactylis (a)44 Festuca (a)45 Bromus (a)46 Brachypodium (a)47 grass – sp. unknown (a)51 mixed grass/herbs (a)61 nettles (a)62 reeds (a)63 Juncus (a)64 bracken (a)71 other (specify)	(b) litter age – all 3 categories (c) litter cover - all 3 categories (f) humus type - both categories LOCATION OF ANIMAL J (a) horizon – all 6 categories (b) position - both categories

Appendix 2

Summary table of species records from each habitat category

The following table shows the number of records for each millipede species from each of the habitat categories used in the analysis of the recording scheme data. The habitat categories are basically the principal habitat groupings (i.e. those with a code number ending in zero) listed on the RA59 card. Records from all of the woodland habitat groupings (codes 090 to 130) have been combined into a single habitat category. Records from the 'Various' principal habitat grouping (code 060) have not been included in the analysis except that records from park /recreation (code 061) and urban / suburban open space (code 069) have been combined and analysed as a single habitat category and records from churchyards (code 068) have also been analysed as a separate habitat category. Records from the 'Other' principal habitat grouping (code 190) have not been included in the analysis.

Species	Building	Cave	Cultivation	Excavation	Grassland	Heath/Moor	Maritime	Rock	Sand dune	Scrubland	Waste ground	Wetland	Woodland	Churchyard	Open space	TOTAL
Adenomeris gibbosa	0	0	1	0	0	0	1	0	0	0	0	0	0	0	0	2
Allajulus nitidus	3	0	17	2	15	0	11	1	0	3	4	0	25	11	7	99
Anamastigona pulchella	0	0	0	0	0	0	0	0	0	0	0	0	5	0	0	5
Anthogona britannica	0	0	0	0	1	0	6	1	0	0	0	0	9	0	0	17
Archiboreoiulus pallidus	7	0	36	3	24	0	10	2	0	4	13	0	110	14	4	227
Blaniulus guttulatus	27	3	190	25	117	7	138	18	5	45	129	2	303	102	38	1149
Boreoiulus tenuis	5	0	42	4	23	2	37	4	1	23	21	1	97	52	13	325
Brachychaeteuma bagnalli	1	0	12	3	7	0	5	2	0	2	2	0	8	4	0	46
Brachychaeteuma bradeae	0	0	9	0	3	0	5	0	0	0	4	0	11	3	0	35
Brachychaeteuma melanops	1	0	18	3	8	0	6	1	0	6	17	2	20	6	4	92
Brachydesmus superus	25	6	121	38	251	16	214	22	6	82	219	37	740	33	20	1830
Brachyiulus pusillus	9	0	52	15	89	1	87	8	16	17	51	14	76	18	11	464
Choneiulus palmatus	16	0	7	3	5	1	20	1	0	3	16	1	6	7	3	89
Chordeuma proximum	0	0	7	9	9	4	31	2	2	8	10	3	180	2	4	271
Chordeuma sylvestre	0	0	0	0	0	0	5	0	0	0	2	0	5	0	0	12
Craspedosoma rawlinsii	0	0	1	1	2	2	13	0	0	1	2	9	52	0	0	83
Cylindroiulus britannicus	20	1	55	15	78	2	78	8	2	42	54	18	299	24	20	716
Cylindroiulus caeruleocinctus	12	0	108	12	75	5	32	0	0	21	59	2	70	57	17	470
Cylindroiulus latestriatus	17	1	13	12	171	34	457	25	192	17	27	11	104	4	7	1092
Cylindroiulus londinensis	0	0	9	1	7	0	5	0	2	1	4	0	27		0	56
Cylindroiulus parisiorum	1	0	8	3	10	0	3	0	0	3	12	0	13	2	0	55

Species	Building	Cave	Cultivation	Excavation	Grassland	Heath/Moor	Maritime	Rock	Sand dune	Scrubland	Waste ground	Wetland	Woodland	Churchyard	Open space	TOTAL
Cylindroiulus punctatus	61	3	190	91	454	106	692	86	25	215	305	71	3093	84	49	5525
Cylindroiulus salicivorus	8	0	0	0	0	0	0	0	0	0	0	0	0	0	0	8
Cylindroiulus truncorum	3	0	2	0	0	0	0	0	0	0	1	0	0	2	0	8
Cylindroiulus vulnerarius	5	0	13	0	0	0	3	0	0	1	5	0	3	1	4	35
Enantiulus armatus	0	0	5	0	1	0	11	0	0	3	1	1	20	0	0	42
Geoglomeris subterranea	2	0	0	9	7	1	4	4	0	2	1	0	35	0	0	65
Glomeris marginata	23	0	66	93	292	31	450	115	10	143	154	20	1544	14	11	2966
Haplopodoiulus spathifer	0	0	4	0	0	0	0	0	0	0	0	0	0	0	1	5
Julus scandinavius	18	0	40	21	120	40	188	21	38	34	44	26	716	10	18	1334
Leptoiulus belgicus	3	0	9	2	5	0	24	1	2	8	13	3	42	0	1	113
Leptoiulus kervillei	0	0	1	2	0	0	25	1	0	5	2	0	47	0	1	84
Macrosternodesmus palicola	8	0	76	5	28	1	21	3	0	15	44	0	94	79	17	391
Melogona gallica	0	0	6	5	1	1	15	1	1	1	4	1	81	5	4	126
Melogona scutellaris	7	0	25	11	23	1	29	3	0	11	13	2	178	10	3	316
Melogona voigti	0	0	0	0	0	0	0	0	0	0	0	0	2	0	0	2
Metaiulus pratensis	0	0	0	0	0	0	0	0	0	0	1	1	0	0	0	2
Nanogona polydesmoides	40	12	137	75	195	16	306	36	10	81	240	13	830	64	24	2079
Nemasoma varicorne	1	0	13	6	58	1	52	4	1	18	8	7	345	6	10	530
Nopoiulus kochii	5	0	0	2	1	0	1	0	0	0	13	0	2	0	2	26
Ommatoiulus sabulosus	20	1	35	22	122	58	296	28	86	32	56	54	480	4	8	1302
Ophiodesmus albonanus	7	0	68	7	21	0	22	1	0	9	26	0	51	38	16	266

Species	Building	Cave	Cultivation	Excavation	Grassland	Heath/Moor	Maritime	Rock	Sand dune	Scrubland	Waste ground	Wetland	Woodland	Churchyard	Open space	TOTAL
Ophyiulus pilosus	52	1	175	62	228	31	290	27	33	89	175	51	1010	49	45	2318
Oxidus gracilis	31	0	1	0	0	0	2	0	0	0	0	0	0	0	0	34
Polydesmus angustus	59	2	145	80	414	123	420	39	34	112	254	57	1381	39	31	3190
Polydesmus barberii	0	0	0	0	0	0	0	0	0	0	0	0	4	2	2	8
Polydesmus denticulatus	4	0	3	3	40	9	47	5	2	11	16	23	199	2	2	366
Polydesmus coriaceus	7	1	123	19	141	4	86	4	4	45	139	12	247	28	14	874
Polydesmus inconstans	9	1	29	11	68	9	74	5	16	11	33	7	81	9	1	364
Propolydesmus testaceus	0	0	1	1	3	0	8	0	0	1	2	0	2	0	0	18
Polyxenus lagurus	18	0	19	3	37	5	92	5	1	3	15	1	46	43	2	290
Polyzonium germanicum	0	0	0	0	1	0	0	0	0	0	0	0	7	0	0	8
Poratia digitata	3	0	0	0	0	0	0	0	0	0	0	0	0	0	0	3
Prosopodesmus panporus	1	0	0	0	0	0	0	0	0	0	0	0	0	0	0	1
Proteroiulus fuscus	18	3	64	27	150	49	296	31	4	87	78	40	1469	26	25	2367
Stosatea italica	1	0	5	1	0	0	10	1	0	0	2	0	15	0	1	36
Tachypodoiulus niger	89	7	257	136	603	93	602	101	45	187	500	57	2130	115	42	4964
Thalassisobates littoralis	0	0	0	0	0	0	21	1	0	0	0	0	0	0	0	22
Trachysphaera lobata	0	0	0	0	0	0	0	0	0	1	0	0	3	0	0	4
Unciger foetidus	0	0	3	0	0	0	3	0	0	0	0	0	0	0	0	6
TOTAL	647	42	2221	843	3908	653	5254	618	538	1403	2791	547	16317	969	482	37233

Appendix 3

Summary tables of species records for each environmental factor

In addition to the principal habitat, the association of millipede species with four other environmental factors was investigated using the recording scheme data. These factors were soil/litter acidity (calcareous or non-calcareous), soil texture (heavy clay/clay or loam or peat or sand/pure sand), coastal/inland bias (less than 15 km from sea or more than 15 km from sea) and synanthropy (rural or suburban/urban bias). The number of records of each millipede species for each of these environmental factors is shown in the following tables.

1. Soil/litter acidity (calcareous or non-calcareous)

Species	Non-calcareous	Calcareous	TOTAL
Adenomeris gibbosa	0	0	0
Allajulus nitidus	8	31	39
Anamastigona pulchella	1	0	1
Anthogona britannica	3	0	3
Archiboreoiulus pallidus	52	82	134
Blaniulus guttulatus	164	338	502
Boreoiulus tenuis	70	99	169
Brachychaeteuma bagnalli	6	22	28
Brachychaeteuma bradeae	8	12	20
Brachychaeteuma melanops	12	38	50
Brachydesmus superus	266	466	732
Brachyiulus pusillus	45	107	152
Choneiulus palmatus	11	11	22
Chordeuma proximum	56	28	84
Chordeuma sylvestre	2	0	2
Craspedosoma rawlinsii	17	4	21
Cylindroiulus britannicus	114	184	298

Cylindroiulus caeruleocinctus	28	154	182
Cylindroiulus latestriatus	104	137	241
Cylindroiulus londinensis	14	11	25
Cylindroiulus parisiorum	5	15	20
Cylindroiulus punctatus	1173	1065	2238
Cylindroiulus salicivorus	0	0	0
Cylindroiulus truncorum	1	0	1
Cylindroiulus vulnerarius	6	17	23
Enantiulus armatus	9	4	13
Geoglomeris subterranea	2	43	45
Glomeris marginata	425	888	1313
Haplopodoiulus spathifer	0	0	0
Julus scandinavius	292	214	506
Leptoiulus belgicus	21	6	27
Leptoiulus kervillei	35	6	41
Macrosternodesmus palicola	56	166	222
Melogona gallica	29	17	46
Melogona scutellaris	82	111	193
Melogona voigti	0	0	0
Metaiulus pratensis	0	0	0
Nanogona polydesmoides	303	464	767
Nemasoma varicorne	110	157	267
Nopoiulus kochii	5	10	15
Ommatoiulus sabulosus	256	187	443
Ophiodesmus albonanus	29	101	130
Ophyiulus pilosus	441	524	965
Oxidus gracilis	1	0	1
Polydesmus angustus	740	615	1355
Polydesmus barberii	0	0	0
Polydesmus denticulatus	67	65	132
Polydesmus coriaceus	90	233	323
Polydesmus inconstans	50	84	134
Propolydesmus testaceus	0	6	6
Polyxenus lagurus	36	63	99
Polyzonium germanicum	0	1	1
Poratia digitata	1	0	1
Prosopodesmus panporus	0	1	1

Proteroiulus fuscus	607	381	988
Stosatea italica	1	11	12
Tachypodoiulus niger	926	1155	2081
Thalassisobates littoralis	2	1	3
Trachysphaera lobata	4	0	4
Unciger foetidus	1	0	1
TOTAL	**6787**	**8335**	**15122**

2. Soil texture (heavy clay/clay, loam, peat or sand/pure sand)

Species	Loam	Peat	Sand	Clay	TOTAL
Adenomeris gibbosa	0	0	0	0	0
Allajulus nitidus	21	0	9	6	36
Anamastigona pulchella	0	0	0	0	0
Anthogona britannica	0	0	0	0	0
Archiboreoiulus pallidus	62	0	3	31	96
Blaniulus guttulatus	285	9	43	143	480
Boreoiulus tenuis	58	3	8	63	132
Brachychaeteuma bagnalli	6	0	1	19	26
Brachychaeteuma bradeae	6	2	2	8	18
Brachychaeteuma melanops	35	1	3	24	63
Brachydesmus superus	335	53	81	304	773
Brachyiulus pusillus	45	12	41	95	193
Choneiulus palmatus	9	2	6	3	20
Chordeuma proximum	53	4	12	28	97
Chordeuma sylvestre	2	0	0	0	2
Craspedosoma rawlinsii	5	11	4	5	25
Cylindroiulus britannicus	149	12	22	100	283
Cylindroiulus caeruleocinctus	136	1	40	47	224
Cylindroiulus latestriatus	57	14	291	20	382
Cylindroiulus londinensis	5	0	8	4	17
Cylindroiulus parisiorum	14	1	4	8	27
Cylindroiulus punctatus	1087	167	268	599	2121
Cylindroiulus salicivorus	0	0	0	0	0
Cylindroiulus truncorum	0	2	0	0	2
Cylindroiulus vulnerarius	10	0	3	6	19
Enantiulus armatus	10	1	3	1	15

Geoglomeris subterranea	14	2	0	6	22
Glomeris marginata	649	49	128	283	1109
Haplopodoiulus spathifer	1	0	0	0	1
Julus scandinavius	242	33	103	103	481
Leptoiulus belgicus	24	0	9	6	39
Leptoiulus kervillei	37	0	1	2	40
Macrosternodesmus palicola	81	2	21	85	189
Melogona gallica	34	1	3	5	43
Melogona scutellaris	52	4	5	48	109
Melogona voigti	0	0	0	0	0
Metaiulus pratensis	1	0	0	0	1
Nanogona polydesmoides	335	29	89	234	687
Nemasoma varicorne	79	14	21	83	197
Nopoiulus kochii	6	0	9	2	17
Ommatoiulus sabulosus	159	75	177	38	449
Ophiodesmus albonanus	74	1	3	48	126
Ophyiulus pilosus	434	63	85	285	867
Oxidus gracilis	1	3	0	0	4
Polydesmus angustus	599	148	213	358	1318
Polydesmus barberii	0	0	0	0	0
Polydesmus denticulatus	51	11	20	60	142
Polydesmus coriaceus	132	15	18	226	391
Polydesmus inconstans	61	8	31	42	142
Propolydesmus testaceus	3	0	0	0	3
Polyxenus lagurus	27	3	11	26	67
Polyzonium germanicum	6	0	0	0	6
Poratia digitata	0	1	0	0	1
Prosopodesmus panporus	0	1	0	0	1
Proteroiulus fuscus	387	102	159	296	944
Stosatea italica	12	0	0	4	16
Tachypodoiulus niger	1015	137	230	556	1938
Thalassisobates littoralis	0	0	4	0	4
Trachysphaera lobata	0	0	4	0	4
Unciger foetidus	1	0	1	0	2
TOTAL	**6907**	**997**	**2197**	**4310**	**14411**

3. Coastal/inland bias
(less than 15 km from sea or more than 15 km from sea)

Species	Coastal	Inland	TOTAL
Adenomeris gibbosa	1	0	1
Allajulus nitidus	31	68	99
Anamastigona pulchella	1	0	1
Anthogona britannica	15	0	15
Archiboreoiulus pallidus	44	180	224
Blaniulus guttulatus	340	744	1084
Boreoiulus tenuis	106	216	322
Brachychaeteuma bagnalli	19	20	39
Brachychaeteuma bradeae	8	27	35
Brachychaeteuma melanops	41	44	85
Brachydesmus superus	529	1168	1697
Brachyiulus pusillus	189	236	425
Choneiulus palmatus	43	40	83
Chordeuma proximum	118	110	228
Chordeuma sylvestre	5	0	5
Craspedosoma rawlinsii	24	52	76
Cylindroiulus britannicus	230	446	676
Cylindroiulus caeruleocinctus	78	387	465
Cylindroiulus latestriatus	700	117	817
Cylindroiulus londinensis	15	43	58
Cylindroiulus parisiorum	3	48	51
Cylindroiulus punctatus	1415	3639	5054
Cylindroiulus salicivorus	8	0	8
Cylindroiulus truncorum	2	6	8
Cylindroiulus vulnerarius	13	26	39
Enantiulus armatus	20	8	28
Geoglomeris subterranea	9	45	54
Glomeris marginata	712	1838	2550
Haplopodoiulus spathifer	0	5	5
Julus scandinavius	424	778	1202
Leptoiulus belgicus	79	11	90
Leptoiulus kervillei	32	19	51
Macrosternodesmus palicola	95	280	375
Melogona gallica	48	60	108

Melogona scutellaris	72	240	312
Melogona voigti	0	0	0
Metaiulus pratensis	0	1	1
Nanogona polydesmoides	550	1252	1802
Nemasoma varicorne	103	412	515
Nopoiulus kochii	13	12	25
Ommatoiulus sabulosus	489	609	1098
Ophiodesmus albonanus	79	170	249
Ophyiulus pilosus	708	1394	2102
Oxidus gracilis	13	19	32
Polydesmus angustus	788	2083	2871
Polydesmus barberii	8	0	8
Polydesmus denticulatus	108	218	326
Polydesmus coriaceus	173 642	815	
Polydesmus inconstans	122	189	311
Propolydesmus testaceus	8	3	11
Polyxenus lagurus	119	158	277
Polyzonium germanicum	0	10	10
Poratia digitata	0	3	3
Prosopodesmus panporus	0	1	1
Proteroiulus fuscus	650	1509	2159
Stosatea italica	11	13	24
Tachypodoiulus niger	1130	3475	4605
Thalassisobates littoralis	23	0	23
Trachysphaera lobata	4	0	4
Unciger foetidus	3	0	3
TOTAL	**10 571**	**23 074**	**33 645**

4. Synanthropy (rural or suburban/urban bias)

Species	**Rural**	**Suburban/urban**	**TOTAL**
Adenomeris gibbosa	0	1	1
Allajulus nitidus	52	41	93
Anamastigona pulchella	1	0	1
Anthogona britannica	12	0	12
Archiboreoiulus pallidus	140	74	214
Blaniulus guttulatus	574	443	1017

Boreoiulus tenuis	142	156	298
Brachychaeteuma bagnalli	16	24	40
Brachychaeteuma bradeae	14	16	30
Brachychaeteuma melanops	35	50	85
Brachydesmus superus	1354	275	1629
Brachyiulus pusillus	294	102	396
Choneiulus palmatus	34	43	77
Chordeuma proximum	186	35	221
Chordeuma sylvestre	5	0	5
Craspedosoma rawlinsii	67	5	72
Cylindroiulus britannicus	446	201	647
Cylindroiulus caeruleocinctus	206	226	432
Cylindroiulus latestriatus	691	94	785
Cylindroiulus londinensis	45	10	55
Cylindroiulus parisiorum	34	16	50
Cylindroiulus punctatus	4149	662	4811
Cylindroiulus salicivorus	0	8	8
Cylindroiulus truncorum	1	4	5
Cylindroiulus vulnerarius	3	35	38
Enantiulus armatus	19	8	27
Geoglomeris subterranea	52	2	54
Glomeris marginata	2164	201	2365
Haplopodoiulus spathifer	0	1	1
Julus scandinavius	943	193	1136
Leptoiulus belgicus	67	16	83
Leptoiulus kervillei	46	5	51
Macrosternodesmus palicola	169	190	359
Melogona gallica	68	31	99
Melogona scutellaris	200	99	299
Melogona voigti	0	0	0
Metaiulus pratensis	0	2	2
Nanogona polydesmoides	1263	421	1684
Nemasoma varicorne	444	44	488
Nopoiulus kochii	4	21	25
Ommatoiulus sabulosus	954	80	1034
Ophiodesmus albonanus	94	137	231
Ophyiulus pilosus	1594	442	2036
Oxidus gracilis	1	23	24

Polydesmus angustus	2224	458	2682
Polydesmus barberii	4,	3	7
Polydesmus denticulatus	288	24	312
Polydesmus coriaceus	572	202	774
Polydesmus inconstans	234	50	284
Propolydesmus testaceus	7	1	8
Polyxenus lagurus	181	67	248
Polyzonium germanicum	9	0	9
Poratia digitata	0	3	3
Prosopodesmus panporus	0	1	1
Proteroiulus fuscus	1770	274	2044
Stosatea italica	15	8	23
Tachypodoiulus niger	3593	730	4323
Thalassisobates littoralis	18	1	19
Trachysphaera lobata	4	0	4
Unciger foetidus	0	3	3
TOTAL	**25 502**	**6262**	**31 764**

References

Alexander, K.N.A. 2003. *Provisional Atlas of the Cantharoidea and Buprestoidea (Coleoptera) of Britain and Ireland.* Huntingdon: Biological Records Centre.

Anderson, R. 1996. *Anamastigona pulchellum* Silvestri (Chordeumatida: Anthroleucosomatidae) an addition to the fauna of northern Europe, recorded from Ireland. *Bulletin of the British Myriapod Group*, **12**, 7-11.

Anon. 1995. *Biodiversity: The UK Steering Group report.* London: HMSO.

Anon. 1998. *UK Biodiversity Action Group, Tranche 2 Action Plans. Volume 1 - vertebrates and vascular plants.* Peterborough: English Nature.

Arthur, D.R., **Cloudsley-Thompson, J.L.** & **Sankey, J.H.P.** 1951. Juniper Hall, Mickleham, Surrey. *Entomologist's Monthly Magazine,* **87**, 275-278.

Baker, A.N. 1974. Some aspects of the economic importance of millipedes. In: *Myriapoda,* edited by J.G. Blower, 621-628. *Symposia of the Zoological Society of London*, No. **32**. London: Academic Press.

Banerjee, B. 1967. Seasonal changes in the distribution of the millipede *Cylindroiulus punctatus* (Leach) in decaying logs and soil. *Journal of Animal Ecology,* **36**, 171-177.

Barber, A.D. 1987. Recent records of *Enantiulus armatus* (Ribaut). *Bulletin of the British Myriapod Group*, **4**, 51.

Barber, A.D. 2003. Fifty years of British myriapod studies: Retrospect. *Bulletin of the British Myriapod and Isopod Group*, **19**, 67-77.

Barber, A.D. 2005. Myriapods from the Eden Project, Cornwall. *British Myriapod and Isopod Group Newsletter,* **11**, 1. Unpublished.

Barber, A.D. & **Blower, J.G.** 1992. S.G.Brade-Birks myriapod papers. *Bulletin of the British Myriapod Group*, **8**, 42-45.

Barber, A.D. & **Fairhurst, C.P.** 1974. A Habitat and Distribution Recording Scheme for Myriapoda and other Invertebrates. In: *Myriapoda,* edited by J.G.Blower, 611-619. *Symposia of the Zoological Society of London,* No. **32**. London: Academic Press.

Barber, A.D. & **Keay, A.N.** 1988. *Provisional atlas of the centipedes of the British Isles*. Huntingdon: Biological Records Centre.

Barber, A.D. & **Mann, D.** 2004. Myriapod papers of R.S.Bagnall, 1889-1962. *Bulletin of the British Myriapod and Isopod Group*, **20**, 13-16.

Barlow, C.A. 1957. A factorial analysis of distribution in three species of diplopods. *Tijdschrift voor Entomologie*, **100**, 349-426.

Berg, M.P. 1995. Preliminary Atlas of the Millipedes of the Netherlands. *Communication European Invertebrate Survey - Nederlands*, **79**, 1-65.

Biernaux, J. 1969. Biotopes à *Schizophyllum sabulosum* L. *Bulletin de la Recherche agronomique de Gembloux*, N.S., **4**, 1–15.

Biernaux, J. & **Baurant, R.** 1964. Observations sur l'hibernation de *Archiboreoiulus pallidus* (Brade-Birks). *Bulletin de l'Institut Agronomique et des Stations de Recherches de Gembloux,* **XXXII** (3), 290–298.

Bilton, D.T. 1990. Recent records of Cylindroiulus vulnerarius (Berlese) (Diplopoda: Julida) in the British Isles. *Bulletin of the British Myriapod and Isopod Group*, **7**, 50–51.

Blower, J.G. 1952. British millipedes with special reference to Yorkshire species. *Naturalist*, **1952**, 145–157.

Blower, J.G. 1958. British Millipedes (Diplopoda). *Synopses of the British Fauna,* No.**11**. London: Linnean Society.

Blower, J.G. 1970. The millipedes of a Cheshire wood. *Journal of Zoology,* **160**, 455–496.

Blower, J.G. 1972a. British Myriapod Group: Inauguration, Field Meetings and News. *Bulletin of the British Myriapod Group*, **1**, 1–4.

Blower, J.G. 1972b. The distribution of British millipedes as known at the end of 1969. *Bulletin of the British Myriapod Group*, **1**, 9–38.

Blower, J.G. 1974. The distribution of British millipedes as known at the end of 1970. In: *Myriapoda,* edited by J.G.Blower, 603–609. *Symposia of the Zoological Society of London,* No. **32**. London: Academic Press.

Blower, J.G. 1979. The millipede faunas of two British limestone woods. In: *Myriapod Biology* edited by M. Camatini, 203–214. London: Academic Press.

Blower, J.G. 1985. Millipedes. *Synopses of the British Fauna (New Series),* No. **35**. London: Linnean Society.

Blower, J.G. 1986. Distribution and variation of the species of *Brachychaeteuma* occurring in Britain and Ireland. *Bulletin of the British Myriapod Group*, **3**, 43–48.

Blower, J.G. & **Fairhurst, C.P.** 1968. Notes on the life-history and ecology of *Tachypodoiulus niger* (Diplopoda, Iulidae) in Britain. *Journal of Zoology,* **156**, 257–271.

Blower, J.G. & **Gabbutt, P.D.** 1964. Studies on the millipedes of a Devon oak wood. *Proceedings of the Zoological Society of London*, **143**, 143–176.

Blower, J.G. & **Miller, P.F.** 1974. The life-cycle and ecology of *Ophyiulus pilosus* (Newport) in Britain. In: *Myriapoda,* edited by J.G.Blower, 503–525. *Symposia of the Zoological Society of London,* No. **32**. London: Academic Press.

Blower, J.G. & **Miller, P.F.** 1977. The life history of the julid millipede *Cylindroiulus nitidus* in a Derbyshire wood. *Journal of Zoology,* **183**, 339–351.

Blower, J.G. & **Rolfe, S.W.** 1956. *Metaiulus pratensis*, gen. et sp. nov., and *Leptoiulus kervillei* (Brolemann), (Diplopoda, Iulidae), from Britain. *Annals and Magazine of Natural History*, (12) **9**, 513–520.

Blower, J.G. & Rundle, A.J. 1980. *Prosopodesmus panporus*, an interesting new species of polydesmoid millipede from the Royal Botanic Gardens, Kew, England. *Myriapodologica*, **1** (4), 27-34.

Blower, J.G. & Rundle, A.J. 1986. *Poratia digitata*, a new British hothouse millipede. *Bulletin of the British Myriapod Group*, **3**, 11-16.

Bocock, K.L., Heath, J. & Blower, J.G.B. 1973. Some observations on the biology of the millipede *Geoglomeris jurassica* Verhoeff, 1915. *Journal of Natural History*, **7**, 691-697.

Bolton, D.E. 1996. *Polydesmus barberii* Latzel and *Anthogona britannica* Mauriès in Devon. *Bulletin of the British Myriapod Group*, **12**, 23-26.

Brade-Birks, S.G. 1920. Notes on Myriapoda XXI: Colobognatha, an order of diplopods new to Britain represented by *Polyzonium germanicum* (Brandt). *Annals and Magazine of Natural History*, (9) **5**, 198-200.

Brade-Birks, S.G. 1930. Notes on Myriapoda XXXIII: The economic status of Diplopoda and Chilopoda and their allies. Part II. *Journal of the South-Eastern Agricultural College, Wye, Kent* , **27**, 103-146.

Brade-Birks, S.G. 1939. Notes on Myriapoda XXXVI: Sources for description and illustration of the British fauna. *Journal of the South-Eastern Agricultural College, Wye, Kent* , **44**, 156-179.

Branquart, E. & Gaspar, C. 1996. Étude comparative des techniques d'échantillonage des macroarthropodes saprophages (Isopoda & Diplopoda). *Acta Myriapodologia. Mémoires du Muséum national d'Histoire naturelle, Paris*, **169**: 485-492.

Bratton, J.H. (ed.) 1991 *British Red Data Books: 3. Invertebrates other than insects*. Peterborough: JNCC.

British Myriapod Group. 1988. *Preliminary atlas of the millipedes of the British Isles.* Huntingdon: BRC.

Brookes, C.H. 1974. The life-cycle of *Proteroiulus fuscus* (Am Stein) and *Isobates varicornis* (Koch) with notes on the anamorphosis of Blaniulidae. In: *Myriapoda*, edited by J.G.Blower, 485-501. *Symposia of the Zoological Society of London*, No. **32**. London: Academic Press.

Brookes, C.H. & Willoughby, J. 1978. An investigation of the ecology and life-history of the millipede *Blaniulus guttulatus* (Bosc.) in a British woodland. *Abhandlungen und Verhandlungen des Naturwissenschaftlichen Vereins in Hamburg*, **21/22**, 104-114.

Causey, N.B. 1943. Studies on the life-history and the ecology of the hothouse millipede *Orthomorpha gracilis* (C.L. Koch, 1847). *American Midland Naturalist*, **29** (3), 670-682.

Corbet, G.B. 1996. *Melogona voigti* (Verhoeff) from the Lothians: A millipede new to the British Isles. *Bulletin of the British Myriapod Group*, **12**, 17-20.

Corbet, G.B. & **Jones, R.E.** 1996. *Haplopodoiulus spathifer* (Brölemann, 1897) in southeast England: A millipede new to the British Isles. *Bulletin of the British Myriapod Group*, **12**, 13-16.

Cotton, M.J. & **Miller, P.F.** 1974. A population of *Cylindroiulus latestriatus* (Curtis) on sand dunes. In: *Myriapoda,* edited by J.G.Blower, 589-602. *Symposia of the Zoological Society of London,* No. **32**. London: Academic Press.

David, J.F. 1984. Le cycle annuel du Diplopode *Microchordeuma gallicum* (Latzel, 1884). *Bulletin de la Société Zoologique de France,* **109** (1), 61-70.

Demange, J.-M. 1958. Trois nouveaux myriapods cavernicoles de France. *Notes Biospéologiques,* **13**, 51-56.

Demange, J.-M. 1965. Remarques sur quelque myriapods cavernicoles de France: Description d'une forme nouvelle du genre *Metaiulus. Bulletin de la Société d'Histoire Naturelle de Toulouse,* **100**, 446-450.

Demange, J.-M. 1970. Myriapodes de Madère et des Acores. *Boletim do Museu Municipal do Funchal,* **25**, 5-43.

Demange, J.-M. 1981. *Les Milles-Pattes.* Paris: Boubée.

Djursvoll, P., **Golovatch, S.I.**, **Johanson, K.A.** & **Meidell, B.** 2001. Phylogenetic relationships within *Polydesmus* sensu lato (Diplopoda: Polydesmidae). In *Progress in Studies on Myriapoda and Onychophora,* edited by J. Wyter and S.I. Golovatch, 37-59. *Fragmenta faunistica,* **43** (*Supplement*).

Doogue, D., **Fairhurst, C.P.**, **Harding, P.T.** & **Jones, R.E.** 1993. A review of Irish millipedes (Diplopoda). In *Biogeography of Ireland: past, present and future,* edited by M.J. Costello and K.S. Kelly, 83-97. *Occasional Publications of the Irish Biogeographical Society,* no.2.

Enghoff, H. 1974. Om tusindbenes udelbredelse i Danmark (Diplopoda). *Entomologiske Meddeleleser.*, **42**, 21-32.

Enghoff, H. 1976a. Morphological comparison of bisexual and parthenogenetic *Polyxenus lagurus* (Linné, 1758) in Denmark and Southern Sweden, with notes on taxonomy, distribution and ecology. *Entomologiske Meddeleleser.*, **44**, 161-182.

Enghoff, H. 1976b. Parthenogenesis and bisexuality in the millipede, *Nemasoma varicorne* C.L. Koch, 1847 (Diplopoda: Blaniulidae). Morphological, ecological and biogeographical aspects. *Videnskabelige Meddelelser fra Dansk Naturhistorik Forening,* **139**, 21-59.

Enghoff, H. 1978. Parthenogenesis and spanandry in millipedes. *Abhandlungen und Verhandlungen des Naturwissenschaftlichen Vereins in Hamburg,* **21/22**, 73-85.

Enghoff, H. 1984. Phylogeny of millipedes - a cladistic analysis. Zeitschrift für *Zoologische Systematik und Evolutionsforschung,* **22**, 8-26.

Enghoff, H. 1987. *Thalassisobates littoralis* (Silvestri) - an amphiatlantic millipede (Diplopoda, Julidae, Nemasomatidae). *Entomologist's Monthly Magazine*, **123**, 205-206.

Enghoff, H. 1991. A revised cladistic analysis and classification of the millipede order Julida. With establishment of four new families and description of a new nemasomatoid genus from Japan. Zeitschrift für *Zoologische Systematik und Evolutionsforschung*, **29**, 241-263.

Enghoff, H. 1994. Geographical parthenogenesis in millipedes (Diplopoda). *Biogeographia*, **70**(1), 25-31.

Enghoff, H. 2005. *Myriapoda*. http://www.zmuc.dk/EntoWeb/kursmyr.doc

Enghoff, H. & Golovatch, S.I. 2003. The millipede genus *Propolydesmus* Verhoeff, 1895 redefined, with a revision of the genus in the Canary Islands (Diplopoda, Polydesmida, Polydesmidae). *Graellsia*, **59**(1), 79-86.

Enghoff, H. & Kime, R.D. 2005. *Diplopoda. Fauna Europaea version 1.2*, http://www.faunaeur.org

Fairhurst, C.P. 1974. The adaptive significance of variations in the life cycles of schizophylline millipedes. In: *Myriapoda,* edited by J.G.Blower, 575-587. *Symposia of the Zoological Society of London,* No. **32**. London: Academic Press.

Fairhurst, C.P. 1984. *British Millipede Survey: 10,000 records for Great Britain. 50 kilometer square maps and habitat information.* University of Salford: Unpublished report for British Myriapod Group.

Fairhurst, C.P. 1985. Interim report on the Millipede Recording Scheme. *Bulletin of the British Myriapod Group*, **2**, 42-47.

Fairhurst, C.P. & Armitage, M.L. 1979. The British Myriapod Survey, 1978. In: *Myriapod Biology* edited by M. Camatini, 183-194. London: Academic Press.

Fairhurst, C.P. & Barber, A.D. 1972. British Myriapod Survey. *Bulletin of the British Myriapod Group*, **1**, 5.

Fairhurst, C.P., Barber, A.D. & Armitage, M.L. 1978. The British Myriapod Survey - April 1975. *Abhandlungen und Verhandlungen des Naturwissenschaftlichen Vereins in Hamburg*, **21/22**, 129-134.

Fairhurst, C.P., Harding, P.T. & Sutton, S.L. 1975. Distribution studies of certain soil invertebrates. In: *European Invertebrate Survey, Proceedings of the Second International Symposium 1973*, edited by J.Heath, 47-55. Huntingdon: ITE.

Fussey, G.D. & Varndell, I.M. 1980. The identification of the sexual form of the bristly millipede, *Polyxenus lagurus* (Linné, 1758) at three coastal sites in England and Wales, using sex ratios. *Naturalist*, **105**, 151-154.

Geoffroy, J-J. 1992. Les Diplopodes edaphiques et souterrains de France: données récentes, répartition des espèces nouvellement décrites et peu connues. *Memoires de Biospeléologie,* **19**, 127-133.

Golovatch, S.I., Gruber, J., Adis, J., Knapinski, S., Zerm, M. & Hansen, B. 2000. Parthenogenetic populations of the millipede *Cylindrodesmus laniger* Schubart, 1945 to be recorded in Europe for the first time (Diplopoda: Polydesmida: Haplodesmidae). *Arthropoda Selecta*, **9** (3), 193–198.

Gregory, S.J. & Campbell, J.M. 1996. An Atlas of Oxfordshire Myriapoda: Diplopoda and Chilopoda. *Oxfordshire Museums Occasional Paper* No. **19**.

Gregory, S.J., Jones, R.E. & Mauriès, J.-P. 1993. A new species of millipede (Myriapoda: Diplopoda: Chordeumatidae) from the British Isles. *Journal of Natural History,* **28**, 47–52.

Haacker, U. 1968. Deskriptive, experimentelle und vergleichende Untersuchungen zur Autökologie rhein-mainischer Diplopoden. *Oecologia,* **1**: 87–129.

Haacker, U. 1970. Experimentelle untersuchungen zur ökologie von *Unciger foetidus* (C.L. Koch). *Bulletin du Muséum d'Histoire Naturelle Paris*, (2), **41** (suppl. 2), 67–71.

Harding, P.T. 1985. Recent records of *Thalassisobates littoralis* (Silvestri). *Bulletin of the British Myriapod Group*, **2**, 6–7.

Harding, P.T. & Jones, R.E. 1994. The occurrence of *Craspedosoma rawlinsii* Leach (Diplopoda) in East Anglia. *Bulletin of the British Myriapod Group*, **10**, 19–21.

Harding, P.T. 1998. Rationale for the inclusion of seven species of millipedes on the UK biodiversity action plan. *Bulletin of the British Myriapod Group*, **14**, 5–9.

Harding, P.T. & Sheail, J. 1992. The Biological Records Centre - a pioneer in data gathering and retrieval. In: *Biological Recording of changes in British wildlife*, edited by P.T. Harding, 5–19. *ITE symposium* no.**25**. London: HMSO.

Harding, P.T. & Sutton, S.L. 1985. *Woodlice in Britain and Ireland: distribution and habitat.* Huntingdon: ITE.

Harper, J. 2004. On some Myriapods new to Wales. *Bulletin of the British Myriapod Group*, **20**, 40–41.

Heath, J., Bocock, K.L. & Mountford, M.D. 1974. The life history of the millipede *Glomeris marginata* (Villers) in north-west England. In: *Myriapoda,* edited by J.G.Blower, 433–462. *Symposia of the Zoological Society of London,* No. **32**. London: Academic Press.

Hoffmann, R.L. 1979. *Classification of the Diplopoda.* Genève: Muséum d'Histoire Naturelle.

Hopkin, S.P. & Blower, J.G.B. 1987. Displaced ocelli in blaniulid millipedes. *Bulletin of the British Myriapod Group*, **4**, 27–29.

IUCN. 2001. *IUCN Red List Categories and Criteria. Version 3.1.* Gland, Switzerland and Cambridge, UK: Species Survival Commission, IUCN,

Jeekel, C.A.W. 1970. *Nomenclator generum et familiarum Diplopodorum;* a list of the genus and family-group names in the class Diplopoda from the 10th edition of Linnaeus, 1758, to the end of 1957. *Monografieen van de Nederlandsche Entomolgische Vereeniging,* Number 5.

Jeekel, C.A.W. 1973. *Julus scandinavius* Latzel, a millipede new to U.S. fauna. *Entomologische Berichten (Amsterdan),* **33**, 101–102.

Jeekel, C.A.W. 2001. The Dutch millipede fauna compared with the British. *Bulletin of the British Myriapod and Isopod Group,* **17**, 43–59.

Johnson, C. 1993. *Provisional atlas of the Cryptophagidae-Atomariinae (Coleoptera) of Britain and Ireland,* edited for the Biological Records Centre by P T Harding and J C M Dring. Huntingdon: Biological Records Centre.

Johnston, G. 1835. A descriptive catalogue of the Insecta Myriapoda found in Berwickshire. *Magazine of Natural History,* **8**, 486–494.

Jones, R.E. 1993a. The vice-county distribution of millipedes in the British Isles. *Bulletin of the British Myriapod Group,* **9**, 25–33.

Jones, R.E. 1993b. Farewell *Mastigonodesmus boncii? British Myriapod Group Newsletter,* **18**, 3. Unpublished.

Jones, R.E. & Keay, A.N. 1986. *Trachysphaera lobata* (Ribaut), a millipede new to Britain, from the Isle of Wight. *Bulletin of the British Myriapod Group,* **3**, 17–20.

Kheirallah, A.M. 1979. Behavioural preference of *Julus scandinavius* (Myriapoda) to different species of leaf litter. *Oikos,* **33**, 466–471.

Kime, R.D. 1978. The Distribution of Millipedes in the South of England - A Preliminary Survey. *Abhandlungen und Verhandlungen des Naturwissenschaftlichen Vereins in Hamburg,* **21/22**, 135–147.

Kime, R.D. 1985. The European Myriapod Survey. *Bulletin of the British Myriapod Group,* **2**, 39–41.

Kime, R.D. 1990a. A provisional Atlas of European Myriapods Part 1. *Fauna Europaea Evertebrata* Vol **1**.

Kime, R.D. 1990b. Spatio-temporal distribution of European millipedes. In: *Proceedings of the 7th International Congress of Myriapodology,* edited by A. Minelli, 367–380. Leiden: E.J.Brill.

Kime, R.D. 1995. Records of millipedes in central southern England. *Bulletin of the British Myriapod Group,* **11**, 37–58.

Kime, R.D. 1997. Year-round pitfall trapping of millipedes in mainly open grassland in Belgium (Diplopoda). *Entomologica Scandinavica, Supplement* **51**, 263–268.

Kime, R.D. 1999. The continental distribution of British and Irish millipedes. *Bulletin of the British Myriapod Group,* **15**, 33–76.

Kime, R.D. 2001. The continental distribution of British and Irish millipedes, part 2. *Bulletin of the British Myriapod and Isopod Group,* **17**, 7–42.

Kime, R.D. 2004. The Belgian Millipede Fauna (Diplopoda). *Bulletin de l'Institut royal des Sciences Naturelles de Belgique, Entomologie,* **74**, 35-68.

Kime, R.D. & **Wauthy, G.** 1984. Aspects of relationships between millipedes, soil texture and temperature in deciduous forests. *Pedobiologia*, **26**, 387-402.

Kime, R.D., **Wauthy, G.**, **Delecour, F.**, **Dufrene, M.** & **Drugmand, D.** 1992. Distribution spatiale et preferences écologiques chez les Diplopodes du sol. *Mémoires de la Société Royal Belge d'Entomologie,* **36,** 661-670.

Leach, W.E. 1814. In: Brewster (ed.) *Edinburgh Encyclopaedia*, **7**, 407-409.

Leach, W.E. 1815. A tabular view of the external characters of four classes of animals, which Linné arranged under Insecta; with the distribution of the genera composing three of three classes into orders, etc and the description of several new genera and species. *Transactions of the Linnean Society of London,* **11**, 306-400.

Lee, P. 2004. *Ommatoiulus rutilans* in Shetland. *British Myriapod and Isopod Group Newsletter,* **8**, 1. Unpublished.

Lee, P. 2005a. A definite new British species from Kew (and Sheffield). *British Myriapod and Isopod Group Newsletter,* **11**, 2. Unpublished.

Lee, P. 2005b. Millipedes from Durham. *British Myriapod and Isopod Group Newsletter,* **11**, 3-4. Unpublished.

Lee, P. 2005c. A *Polydesmus* hot spot in Suffolk. *British Myriapod and Isopod Group Newsletter,* **11**, 4. Unpublished.

Lee, P., **Gregory, S.J.**, **Keay, A.N.** & **Read H.J.** 2005. Report on a survey of the pill millipede *Trachysphaera lobata* (Ribaut, 1954) in the Bembridge area of the Isle of Wight. Unpublished report for English Nature.

Lindroth, C.H. 1957. *The faunal connections between Europe and North America.* Stockholm: Almquist & Wiksell.

Lohmander, H. 1925. Sveriges Diplopoder. *Götesborgs Kungliga Vetenskaps-och Vitterhets Samhälles Handlingar*, **30**, 1-115.

Lohmander, H. 1956. Faunistiskt fältarbete 1955 (Huvudsakligen Södra Värmland). *Götesborgs Naturhistoriska Museum Ärstyck,***1956**, 32-94.

Luff, M.L. 1998. *Provisional atlas of the ground beetles (Coleoptera, Carabidae) of Britain.* Huntingdon: Biological Records Centre.

Mauriès, J.-P. 1960. Diplopodes de la Région Toulousaine. *Bulletin de la Société d'Histoire Naturelle de Toulouse,* **95**, 100-104.

Mauriès, J.-P. 1965. Une nouvelle sous-espèce cavernicole de *Metaiulus pratensis* Blower & Rolfe, 1956. *Annales de Spéléologie*, **20**, 519-522.

Meidell, B.A. 1968. *Microchordeuma gallicum* (Latzel, 1884) new to Scandinavia. *Norsk Entomologisk Tidsskrift,* **15**, 41-46.

Minelli, A. 2003. Diplopoda. In: *Checklist of the species of the Italian fauna version 2.0,* http://www.faunaitalia.it/checklist/invertebrates/Diplopoda.html# Diplopoda , edited by F. Stoch.

Morgan, I.K. 1988. Recent records of Myriapods in South-West Wales. *Bulletin of the British Myriapod Group*, **5**, 11-23.

Pedroli-Christen, A. 1977. Etude des Diplopodes d'une tourbière du Haut-Jura. *Bulletin de la Société Neuchâteloise des Sciences Naturelles,* **100,** 21-34.

Pedroli-Christen, A. 1993. Faunistique des Mille-pattes de Suisse (Diplopoda). *Documenta faunistica Helvetiae14.* Neuchatel: Centre Suisse de Cartographie de la Faune.

Pierrard, G. & Biernaux, J. 1974. Note à propos des diplopodes nuisibles aux cultures tempérées et tropicales. In: *Myriapoda,* edited by J.G.Blower, 629-643. *Symposia of the Zoological Society of London,* No. **32**. London: Academic Press.

Pierrard, G., **Bonte, E.** & **Baurant, R.** 1963. Observations sur l'hibernation de *Blaniulus guttulatus* Bosc. *Bulletin de l'Institut Agronomique et des Stations de Recherches de Gembloux,* **XXXI** (1), 127-141.

Pocock, R.I. 1893. Notes upon some Irish Myriapoda. *Irish Naturalist*, **2**, 309-312.

Pocock, R.I. 1903. Myriapoda in *Victoria History, County of Essex* **1**, 193-195.

Rantala, M. 1974. Sex ratio and periodomorphosis of *Proteroiulus fuscus* (Am Stein). In: *Myriapoda,* edited by J.G.Blower, 463-469. *Symposia of the Zoological Society of London,* No. **32**. London: Academic Press.

Read, H.J. 1988. The life history of *Chordeuma proximum* Ribaut from a wood in Avon. *Bulletin of the British Myriapod Group,* **5**, 24-27.

Read, H.J. 2002. Obituary. John Gordon Blower 1923-2001. *Bulletin of the British Myriapod and Isopod Group,* **18**, 5-12.

Read, H.J., **Corbet, G.** & **Jones, R.E.** 2002. *Cylindroiulus salicivorus* Verhoeff, 1908: A millipede new to Britain. *Bulletin of the British Myriapod Group,* **18**, 41-44.

Remy, P. & Hoffman, J. 1959. Faune des Myriapodes du Grand-Duché de Luxembourg. *Archives de la Section des Sciences de l'Institut Grand-Ducal,* **26**, 199-236.

Richards, J.P. 2003. J. Gordon Blower as inspiration for local distribution mapping of millipedes: a personal view. *Bulletin of the British Myriapod and Isopod Group,* **19**, 17-19.

Schömann, K. 1956. Zur biologie von *Polyxenus lagurus* (L, 1758). *Zoologische Jahrbücher (Systematik),* **84**, 195-256.

Schubart, O. 1930. *Brachychaeteuma verhoeffi* nov. spec., ein neuer deutscher Diplopode. *Zoologischer Anzeiger,* **90**, 38-45.

Schubart, O. 1934. Diplopoda. *Tierwelt Deutschlands.* **28**, 1-318.

Schubart, O. 1938. Ein für Deutschland neuer Diplopode aus westdeutschen Höhlen. *Sonderdruck aus den Mitteilungen uber Höhlen und Karstforschung,* **1938**, 133-137.

Schubart, O. 1946. Uma segunda especie do genero *Cylindroiulus* (Diplopoda) encontrada no Brasil. *Communicaciones Zoologicas del Museo de Historia Natural de Montevideo,* **29** (II), 1-5.

Schubart, O. & Husson, R. 1937. Les diplopodes de cavités souterraines du Nord-Est de la France. *Bulletin de la Société Zoologique de France,* **61**, 484-502.

Snider, R.M. 1981. The reproductive biology of *Polydesmus inconstans* (Diplopoda: Polydesmidae) at constant temperatures. *Pedobiologia*, **22**, 354-465.

Spelda, J. 1999. Verbreitungsmuster und Taxonomie der Chilopoda und Diplopoda Südwestdeutschlands. Unpublished dissertation, University of Ulm.

Stephenson, J.W. 1961. The biology of *Brachydesmus superus* (Latzel). *Annals and Magazine of Natural History.* (13) **3**, 311-319.

Templeton, R. 1836. Catalogue of Irish Crustacea, Myriapoda, and Arachnida, selected from the papers of the late John Templeton, Esq. *Magazine of Natural History*, **9**, 9-14.

Turk, F.A. 1944. Myriapoda (Chilopoda and Diplopoda) from Cornwall with notes and descriptions of forms new to the British fauna. *Annals and Magazine of Natural History.* (11) **11**, 532-551.

Van Den Haute, L. 1999. Faunistiek en ecologie van Diplopoda (Myriapoda) in Vlaamse bossen. Unpublished thesis, University of Ghent.

Verhoeff, K.W. 1908. Über Diplopoden 10(30). Aufstaz: Zur kenntnis der Juliden und über einige Polydesmiden. *Archiv für Naturgeschichte,* **73**, 423-474.

Verhoeff, K.W. 1928. Diplopoda I. *Bronn's Klassen und Ordnungen des Tierreichs,* **5** (2), 1-1072.

Zulka, K.P. 1996. Submersion tolerance of some diplopod species. *Acta Myriapodologica. Mémoires du Muséum national d'Histoire Naturelle,* **169**, 477-481.

Species index